PAPER — ART & TECHNOLOGY

PAPER — ART & TECHNOLOGY

Based on presentations given at
the International Paper Conference
held in San Francisco, March 1978

Paulette Long, Editor
Robert Levering, Associate Editor

World Print Council
San Francisco
1979

Acknowledgements

This book was supported by a grant from the National Endowment for the Arts in Washington, D.C., a federal agency, with substantial support from the California College of Arts and Crafts, Oakland, and the Simpson Paper Company, San Francisco.

We would also like to thank the International Paper Company Foundation, Andrews/Nelson/Whitehead, and Special Papers, Inc., who have helped assure that we could distribute copies of the book to printmaking departments around the country.

Many thanks also the the WPC staff whose constant encouragement helped the editors through much stormy weather, and especially to Jim Felici who volunteered his sharp editorial skills to help shape the final product.

Cover Art:
Charles Hilger
Discovery Series (Bow) 1978
Vacuum-formed paper
48″ × 30″ × 11″
Courtesy of the Smith-Andersen Gallery, Palo Alto, California

Designer: Brenn Lea Pearson
Printed on Shasta Suede from the Simpson Paper Company, by Mastercraft Press, Inc.

Library of Congress Cataloguing in Publication Data
Main entry under title:
Paper — art & technology.
    1. Papermaking and trade — History — Addresses, essays, lectures. 2. Paper — Addresses, essays, lectures.
I. Long, Paulette.
TS1090.P36   676'.2   79-10879
ISBN 0-9602496-0-5

FOREWORD

Paper—Art & Technology is a book based upon the material presented at an international conference of the same name presented by the World Print Council in March 1978 in San Francisco.

A large number of respected experts, from scientists to artists, have developed a comprehensive series of essays dealing with the theme of paper as it has been and as it is currently being used by the creative community. The resulting publication becomes a unique resource for technological and aesthetic information concerning the long historical relationship between paper and art both as support and medium.

The conference and book were formed at a time when interest in paper was sweeping the art world. It was a time when, with rare exceptions, the process of working with paper and paper pulp seemed to overwhelm the artistic vision. However, it is interesting to note that in the one year that has passed since the conference, papermaking and paper use by artists

have reached a new plateau of sophistication. Recent projects which have brought together the combined skills of a master paper craftsman and an artist of distinction have resulted in unique objects that expand the art horizon. I speak of David Hockney's collaboration with Kenneth Tyler and the working relationship between Sam Francis and Garner Tullis, among others. The results so completely integrate the paper process and the artist's intent that it is as if a new form had been produced.

One hopes that this excellent compendium, *Paper—Art & Technology*, will develop a wide audience for the purpose of continued art production, appreciation, and understanding.

Henry T. Hopkins, Director
San Francisco Museum of Modern Art
February 1979

INTRODUCTION

Paper has been with us for nearly 2,000 years. It is one of the simplest and most beautiful materials known to man, yet it is also one of the least appreciated. Until the invention of the Fourdrinier papermaking machine in the late 18th century, paper was a highly treasured commodity with each sheet painstakingly made by hand. The refined Fourdrinier machine allows a modern-day papermill to produce a 14-foot-wide ribbon of paper at the rate of 1,700 feet per minute. With this astounding production capability, we are literally surrounded today by a multitude of paper objects and products. It is no wonder that most of us have become blasé about this once-treasured substance.

The artist, on the other hand, has always appreciated the finely made sheet of paper. The recent renaissance of fine papermaking means the contemporary artist now has a great variety of papers produced by hand, mold, or machine that can be custom-made for almost any printmaking or other artistic use. As the craft of hand papermaking is being revived in small mills throughout the world, artists are becoming increasingly involved with paper as a substrate and medium for art work. It is this renaissance of papermaking that led World Print Council to organize a large international conference, "Paper — Art & Technology," and encouraged the publication of this book.

Preparations for this book, and the conference on which it is based, began in the early winter months of 1977. Our exhibition, WORLD PRINT '77, had completed a successful showing at the San Francisco Museum of Modern Art and was safely packed and on its way to tour the country with the Smithsonian Institution Traveling Exhibition Service. Having established contact with thousands of print artists throughout the world, the World Print Council felt it should offer services, other than the exhibition, to its constituents. The artists had conveyed their interest in the resurgence of the papermaking craft, and the art world was becoming increasingly involved in paper as art. A highly developed communications network gave WPC the facility to bring together experts from the arts and industry alike to examine the aesthetic and technical potential of this rapidly expanding field. A conference seemed to be the perfect vehicle for this forum.

We were aware of three previous paper conferences held in the United States since 1975. Papermaker Joe Wilfer organized one for other hand papermakers in Appleton, Wisconsin, in 1975. At that conference, the Friends of Paper, a very small group of dedicated papermakers, met to exchange their knowledge of the craft. A somewhat larger conference was held in 1976 at the Santa Barbara Museum in conjunction with the exhibition, "The Handmade Paper Object," curated by Richard Kubiak. In October 1977, more than 100 papermakers, artists, and scholars attended a third paper conference at the Center for the Book Arts, in New York City.

The interest shown in these conferences as well as the growing demand for fine papers by individual artists and large commercial printmaking studios convinced us that there was a growing desire for a comprehensive, international conference on paper. Following long hours of discussion with our many friends, including Don Farnsworth and Bob Serpa, two fine local papermakers, as well as a multitude of print and paper artists, museum personnel, and publishers, WPC decided to proceed with the funding search and planning to make the conference a reality.

"Paper — Art & Technology" was held March 23-25, 1978, at the San Francisco Museum of Modern Art and attracted more than 450 participants from as far away as Japan, Sweden, and Tanzania. As reflected in this book, the conference participants spent three days examining these topics: History and Technique; Technology; and Art: Paper as Substrate and Medium. Eighteen renowned speakers from a variety of fields participated in the lectures and panel discussions, and they presented a substantial quantity of primary source material.

WPC felt that the quality of this information called for wider distribution than among the conferees, and at the suggestion of Kenneth E. Tyler obtained a grant from the National Endowment for the Arts to document the information in the form of a book. With few exceptions, the following essays are based on transcripts of the conference presentations.

The lectures have been edited to ease the transition from the spoken word to print, and the editors have attempted to maintain the freshness of the live presentations. Although many of the visual aids used at the conference could not be reproduced, we have included illustrations wherever possible.

Both the conference and the book required a great deal of cooperation and support, and the World Print Council wishes to extend its appreciation to the following people and organizations: the staff of the California College of Arts and Crafts gave immeasurable support and counsel throughout the project; the San Francisco Museum of Modern Art generously provided space for the conference; the Japan Foundation funded the journeys of Yasuichi and Akira Kubota from Japan, enabling them to demonstrate the magical technique of Japanese papermaking; partial funding for the conference came from the Simpson Foundation and the following paper companies — Arjomari-Prioux, Andrews/Nelson/Whitehead, Rising, and Strathmore; seventeen local museums and galleries offered an additional dimension by presenting satellite exhibitions relating to paper as art. We are also most grateful to Robert J. Seidl, president of the Simpson Paper Company, for his generous and early support of the project. His enthusiastic encouragement carried us through the ten long months of planning and fundraising.

Again, WPC wishes to extend appreciation to all those who made the conference a great success — speakers, participants, and countless volunteers. Thanks also to the staff and consultants who worked long and hard to document the information and develop this book, and to the National Endowment for the Arts, the California College of Arts and Crafts, Simpson Paper Company, the International Paper Company Foundation, Andrews/Nelson/Whitehead, and Special Papers, Inc., which have enabled us to distribute this book without charge to colleges and universities throughout the United States.

As Robert J. Seidl so appropriately stated in his opening remarks at the conference, "We come here from many backgrounds. It is the fiber we use that binds us together in interest and has the great potential to shape our creative directions and to enrich our lives in the process." I hope this book will also serve to enrich our awareness and understanding of this simple yet complex, common yet beautiful material — paper.

Trisha Garlock, Director
World Print Council

TABLE OF CONTENTS

A HISTORY OF PAPER

Leonard B. Schlosser

Contemporary hand papermaking is going in a number of directions, some of which are completely new art forms. From the point of view of technique, however, not much hasn't been employed before.

I will attempt to show, with the use of books and illustrations as references, a view of the history of paper as a graphic medium, as well as integrating that history with paper's use in the printing process. Most of the books I shall refer to are from my own personal collection and are not available in ordinary libraries. Some are very rare, while almost all are uncommon.

This will not be a complete history of paper because I think that would probably take a year's course (I know because I used to teach one). Rather, this will be a review of high spots and landmarks in the development of paper, mostly before 1860. Since paper produced before that time was primarily handmade paper, this subject relates directly to an increasing interest among artists and others today.

Let me begin with a relevant quotation from a book by historian Jean Gimpel, published last year, called *The Medieval Machine and the Industrial Revolution of the 13th Century*. There *was* an industrial revolution in the 13th century, and much of it concerned, interestingly enough, the development of mills. One of those kinds of mills driven by water power was the first, the earliest, of European paper mills. Gimpel

wrote, "Our ignorance of the history of technology prohibits us from understanding fully the economic and political evolution of our time. It also distorts our picture of the past. We are convinced that we are living in the first truly technological society in history, and looking around we can see only continuous progress in technology and science. The historian of technology must correct this belief. Our Western civilization has reached a technological plateau that will extend well into the third millenium."

As I develop my theme, I will attempt to refer to certain plateaus that have been reached in earlier times in the graphic arts and, specifically, in paper, and I will hope to demonstrate the aptness of this quotation. The development of paper for writing and printing is essentially one of the development of a surface for the acceptance of a communications medium. Throughout its history and right up to the present moment, paper has always been a material, at least in a communications sense, which has been employed as a substrate for writing, printing, or some other method of communication.

Of course, I leave out the great developments in packaging that began to take place in our own time. Because paper is and has been a substrate for communications media, its *refinement* of surface in order to carry images has always been extremely important. Many of the developments that have recently taken place in using paper as structural object, as part of a print or as part of an art form, are antithetical to that history. That doesn't mean that they're bad, but the history of paper does not lead us in that direction, which I should mention at this point.

Forerunners to Paper

Paper was preceded by a number of flexible predecessors—palm leaves and papyrus being the two principal ones. It had rigid predecessors, of course, as well—wax-coated tablets of wood, clay used for cuneiform writing, stone, and a whole variety of other materials. But with the development of the earliest *flexible* substrates, it became possible to make writing portable in lightweight form and to communicate from one place to another readily instead of being weighed down with clay tablets or some other heavy material.

The development of paper goes back to the Orient. The date is often cited as A.D. 105. Now paper didn't just happen in A.D. 105. It has been demonstrated by recent scientific exploration that paper and what is sometimes called "proto-paper" (forerunners made of silk and other materials) existed before 105. What is significant about the date A.D. 105 is that in that year it is recorded in history that a Chinese gentleman by the name of Ts'ai Lun, the chief eunuch in the court of the then emperor, arrogated to himself the patents for making paper, which probably already existed at that time. He's really, therefore, likely not the inventor of paper but is the first to record that paper existed.

The so-called *Million Charms* of the Empress Shotoku is an example of early Oriental paper used for printing. It is a piece of paper in the form of a small scroll about 1½ inches high. It is also probably the second oldest extant printing of text material on paper and was printed in the year A.D. 770. That is recorded fact.

There were a number of these (no one really knows how many, but the number is always given as a million) placed in wooden pyramids and deposited in four different temples in Japan. These are all prayers that were given out by the Empress Shotoku in order to thank the gods for the deliverance of the then country of Japan from a plague.

Relatively few of these have survived, although one of the four temples is still standing, the one at Horyu. A few of the charms (*dharani*) were sold in the 1890s in order to raise funds for the temples and have found their way into libraries and private collections. This was a great ecumenical work in many ways because it is a Buddhist prayer written in the Japanese language, transliterated from Sanskrit, and printed in Chinese characters, probably from stone.

I spoke of papyrus before. Pliny's *Natural History,* which in manuscript form was written in the first century A.D. and printed for the first time in 1469 in Venice, is often cited as having mentioned paper, called *charta* in Latin. In point of fact, Pliny referred to papyrus rather than paper because there was no paper in the first century A.D. in Europe. The first illustrated Pliny was made in 1513 (*Fig. 1*), and it has an illustration at the head of *Book 13* of papyrus-making as Pliny visualized it taking place.

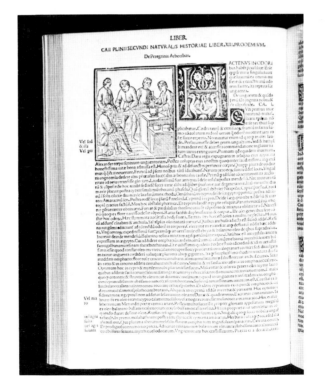

1. Caius Plinius Secundus. Historia Naturalis. Venice. Sessa. 1513. 11 5/8" x 8¼". The first illustrated edition. (Illustrations in this chapter courtesy of Leonard B. Schlosser.)

Unlike paper, papyrus is not a mat of fibers but is a series of split reeds that have been laid across one another at right angles and then pounded with a mallet to permit their natural juices and close contact to hold them together. This is not an unusual product although there is relatively little of it made now, for the papyrus reed has virtually disappeared from Egypt, its place of origin, and its present interest is in large part historical.

From China to Europe

Paper, as I mentioned earlier, originated in China and journeyed from East to West. It journeyed from the Chinese Empire (actually samples have been found of paper as old as the first century A.D. in Western Turkestan) by way of Samarkand, whence the Arabs carried it to North Africa, and then to Europe. In the capture of Samarkand in 751, the Arabs captured, so the story goes, among their Chinese prisoners a number of people who were papermakers, and they extorted or extracted from them (one suspects none too gently) the secrets of their art. The Arabs took the papermaking secrets with them back to North Africa, and they began making paper in North Africa (Arabic paper prior to the year 1000 exists, much of it in the Royal Library in Vienna where there is a major collection of such material).

Finally paper went from North Africa to Europe by way of Spain. The Moors apparently took paper with them in their conquest of the Iberian Peninsula, where the first mill was established on the continent of Europe in about 1100 in the town of Xativa in Spain. Paper went from Spain to Italy, where in 1275-76 the first mill was established in Fabriano, which is still a papermaking town. From there, paper went to France in the mid-14th century (around 1348, as the records indicate), to Germany before 1390, and then spread across the rest of Europe.

It is important that we recognize what happened in the transition of paper from East to West. In the first place, there is nothing that we know of published in the Orient in printed form about papermaking until almost the end of the 18th century. There are contemporary manuscripts that exist, but the printed works that cover the process are very few and far between until after the end of the 18th century, the first being

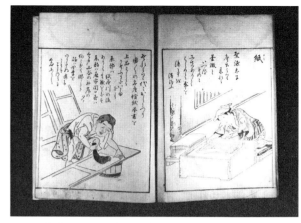

2. *Tachibana Minko.* Saiga Shokonin Burui. *Edo. 1784. 2nd edition. 10¾ " x 7½ ". Classified Artisans in colored pictures.*

the great Japanese papermaking manual *Kamisuki Chohoki,* done in 1798. There is, however, a single picture of Oriental papermaking as part of a *Book of Trades* by the Japanese artist Tachibana Minko, made in 1784 (*Fig. 2*).

Oriental paper was, of course, made for writing because there was relatively little printing, although there was some in xylographic form in the early days. The paper had a characteristically soft surface because it was written on with a brush, and it was made differently and out of different materials from even the earliest Western paper. The Japanese, Chinese, and Koreans were using wood as a papermaking material long before it was ever used in the West.

Papermaking had to undergo a change that converted it from an Eastern writing material to a Western material able to take the water-based writing inks made with ox gall and lampblack that had a somewhat acidic nature and permitted the use of a quill pen without scratching up the surface. Techniques, therefore, were developed in making European paper that made it quite different from the paper made in the East. European paper, for example, was characteristically made out of rags rather than raw vegetable fiber. Sheets of European paper were sized fairly early in the game with glue or gelatin, as opposed to the vegetable "mucilage," which was used to size fiber for Oriental paper. The European papermaking mould, typically, is a fixed wooden mould rather than one with the flexible cover used in the Orient. Finally, from earliest times, European paper normally was dried in the air, usually by being hung over horsehair ropes rather than in contact with a board as is done in Japanese papermaking. European paper *required* a different surface and was, therefore, made differently.

There was some early printing in the Far East, as I earlier mentioned. Illustrative printing was far more common in the Orient in the form of xylography than it was in Europe because there was virtually no printing of illustration in Europe until the blockbooks and single sheet prints of the mid-1400s.

Early European Papermaking

Let me describe to you how early 14th century paper was constructed. One notes the very coarse laid lines and the wide spacing of the chain lines. This is typical of the materials of the time because wire on a papermaking mould was not drawn wire (the technique of drawing wire had not yet been refined) but was made from sheets of copper or bronze. It was cut up very fine, beaten with a hand hammer into wire form, and the resulting product was, therefore, coarse. The wires were laid, literally, onto the surface of the mould, then sewn to the ribs of the mould, giving rise to both laid lines and chain lines (the latter from the sewing to the ribs) in paper formed on the mould. Typically in 14th century paper (about the earliest European paper that has survived, although there's a little that is earlier), we see these very coarse lines because of the nature of the wire.

Watermarks were used early. They are thought to have been used to identify the papermaker, though no one is really sure. There are some of them that have religious significance, such as a Gothic "M" surmounted by a cross. However, it's entirely possible that the papermaker's name began with an "M," and the double cross, the cross of Lorraine, is a religious usage.

The earliest extant picture of a papermaker appears in the book the *Panoplia* or Book of Trades, by a German named Hartmann Schopper, and illustrated by the Swiss artist Jost Amman, published in Frankfurt in 1568 (*Fig. 3*). This picture has become justly famous. It is one of 139 different trades and professions shown in the book. It is probably not the earliest picture of papermaking ever made, but it is simply the oldest extant. One should notice the mould in the hands of the papermaker in this picture. The chain lines are depicted very clearly and, as is true in modern hand papermaking practice and in European practice from earliest times, the chain lines run in the direction perpendicular to the long dimension of the mould. There are some technical reasons for that which have to do with release of the sheet from the mould when it is couched. I don't want to get into that for the moment, but there is a logical reason for that convention.

In the background of this picture, one can notice a device called a stamping mill (or stamper), consisting of heavy pieces of timber, which move up and down in a vertical direction and which were used to defiber pulp in water. The stamper is driven by a waterwheel, which may be seen outside the window. The only thing missing from this picture is the operation of couching, which, strangely enough, does not appear in most early European pictures of papermaking.

The stamping mill was the first major improvement made by the West in the art of papermaking. So far as we know, in the Orient and in North Africa, pulp was prepared by defibering rags by pounding them up and down in a large hand-operated mortar and pestle. The stamping mill was invented either in Spain or in Italy in earliest times, but let it suffice to say that it was in use by the late 13th century.

In the Jost Amman book of 1568 (incidentally the book occurs both in the vernacular, German, as well as in Latin), there is an illustration of a printer, a *Buchdrucker,* or, in Latin, *typographus* (*Fig. 4*). Notice that the use of the printing press and its particular technique is already pretty well established. This is, after all, more than one hundred years after the invention of printing. By this time, the making of paper was in full swing in Europe. It was considered something of a secret, handed down from master to apprentice, father to son, and there are no published manuals on the making of paper in Europe until mid-18th century. There were some things printed earlier that were not intended as manuals but

3. *Illustration by Jost Amman from Hartmann Schopper's* Panoplia omnium...artium... *Frankfurt a/M. Feyerabend. 1568. 6 1/8" x 3¾". Earliest surviving illustration of a papermaker in the West.*

4. *Illustration by Jost Amman from Hartmann Schopper's* Panoplia omnium...artium.. *Frankfurt a/M. Feyerabend. 1568. 6 1/8" x 3¾". The Printer.*

were intended as engineering books, and there is absolutely nothing in the way of printed manuals of the art (at least that has survived) before that time.

A most interesting book was printed in Ingolstadt in 1579, and it tells us very clearly that even this early there were distinctions being made between writing and printing paper. As stated in the book, "The thick and hard paper, which results when the rags are not rotted enough, is the best printing paper but the poorest writing paper." There were, as you see, already distinctions being made between printing and writing paper, and these distinctions occur more and more as illustrational processes begin to advance.

About 1550, the woodblock was beginning to be replaced for illustration by the engraving. The engraving presented a problem to the printer, however, because it had to be printed separately on a rolling press. The advantage of woodblocks was that they could be locked up with type and printed on the paper in one impression. This was not the case with engravings or etchings, which required the use of a rolling press.

As far as we know, the earliest illustration of a rolling press is contained in an Italian machinery book of 1607 called *Teatro Nuovo de Machine ed edifici,* written by Vittorio Zonca and published in Padua (*Fig. 5*). It also contains the first technical illustration of a paper mill, a stamping mill (*Fig. 6*). You may note that I consider this technical, whereas the Jost Amman illustration I referred to earlier is not, strictly speaking, technical. The reason is that plates like those in the Zonca book were used, believe it or not, as the blueprints of the time. Mill builders had no reference points unless they were able to go to a printed work of some sort in order to reproduce a mill or were able to draw another one firsthand. Within the limited resources of the artists and draftsmen of the time, therefore, these plates were cut as engravings and printed. Consequently, one of the reasons that machinery books have survived in relatively small numbers is that they were cut up and used by mill builders literally as blueprints.

In Zonca's picture of the stamping mill, you see the water wheel driving it, in this case, an undershot water wheel. The water wheel turned the line shaft, which had lugs on it, like the lugs on a music box shaft, which in turn cammed these individual pilings or hammers up and down. The fibrous material

5. *Vittorio Zonca.* Teatro nuovo de machine ed edifici. *Padua. 1607. 11 1/16" x 7¾". Probably the earliest illustration of a rolling press, from the first edition of this machinery book.*

(rags in small pieces and water) were in the trough as shown. The stone trough is pictured taken apart as well, while the hammer and iron shoeing for the foot of the hammer may also be seen.

Iron begins to contaminate paper relatively early because of iron shod stampers that were sometimes but not always used.

CARTIERA OVERO PISTOGIO CHE PESTA LE STRAZZE PER FAR LA CARTA.

6. Vittorio Zonca. Teatro nuovo de machine ed edifici. *Padua. 1607. 11 1/16" x 7¾". First edition of the book showing the earliest technical depiction of a stamping mill.*

The action of the stamper is quite different from that of the Hollander beater, which we'll see later, because it bruises the rag fiber. It takes apart the fibers by vertical pounding action rather than by the mechanical cutting and separation that the Hollander beater performs at the same time that it bruises the fibers.

Stampers produce very good paper because the fiber is heavily hydrated by the action of the stampers, depending on how long it's left in. Rags were not simply ripped up into pieces and put into stampers. Instead, they were first cut up into pieces, left on wet stone floors to rot somewhat (a process called "retting") in order to let the natural resins of the fibers weaken sufficiently to permit the rags to be defibered when put under the hammers of the stamping mill.

Early Materials for Paper

Papermaking was growing all of this time, while raw materials were declining in quality, possibly because of larger demand, or possibly because people were beginning to wear cotton clothing and therefore provided castoffs of cotton rags for papermaking. Linen was the common material for clothing until the advent of overseas possessions and the use of cotton in the 17th century. Before that time, shirts, nightgowns, sheets, and whatever else people wore in the way of woven cloth was in the main made from linen, so that early paper in Europe is mostly linen paper. There was a good deal of hemp, often from waste rope, used as well. There was sometimes cotton, but later on cotton came more and more into use until by the mid-18th century, it became very common.

One of the 18th century parlor games played by the intelligentsia who were interested in papermaking, was to attempt to determine the differing origins of *charta linea* (linen paper) and *charta cottonea* or *bombycina*, which is cotton paper. The origins were relatively more simple than the people then were able to determine because they had to deal really with the use of fiber for cloth.

Early illustrations tell us a lot about the customs of the time. At some point or other many of you will surely have read Dard Hunter's book, *Papermaking, The History and Technique of an Ancient Craft* or his *Papermaking Through Eighteen Centuries.* Hunter makes much of the illustration from the Nuremberg Chronicle of the city of Nuremberg with a mill shown outside the walls. He claims that the mill is that of Ulman Stromer, who was known to have been making paper in the Nuremberg area in 1390.

There is no doubt that the cut is of the city of Nuremberg,

and undoubtedly the structure shown in the foreground is a mill. We have nothing, however, that tells us that it is Stromer's mill, the first paper mill in Germany, and therefore the earliest depiction of a paper mill. That is a lovely story, but I'm afraid it's mythical.

Engraving, Etching, and Mezzotint

New illustrative methods began to come along at about this time. I spoke of engraving, which had begun in the mid-15th century but had not begun to displace the woodblock until the early 16th. Etching had also begun to appear. Etching was used in a crude form in the early 16th century and was used very elaborately by the beginning of the 17th, notably in the hands of the first great adventurer with etching techniques, Jacques Callot, in France and Italy.

In the year 1662, a brief description of mezzotint was published in John Evelyn's book, *Sculptura*. By 1662 we have not only the woodblock used to print illustrations (and we have, of course, printing of type), but we have etching, engraving, and mezzotint. Mezzotint was also called the "dark manner" (*manière noire*) or the English manner (*manière anglaise*) because, as you know, in mezzotint you work from dark to light instead of light to dark. All four of those reproductive methods, three of them intaglio and one relief, were in full swing by this time. There are no further major changes in the use of illustrational methods until the late 18th century.

The significant thing about illustrational processes is that they require paper of a different surface from typographic printing. It is very apparent in looking through finely printed books that used engravings or etchings as illustrations that the paper on which the engravings or etchings are printed is of a different nature, often thinner, always with a more refined surface than that on which typographic material is printed. If the two are printed together, of course, they use the same paper, in which case the printer attempted to use refined paper. The point I wish to make is that there is strong differentiation, perhaps not to the modern eye, but in use then, between writing paper and paper used for printing or illustrational purposes.

Papermaking Comes to England and America

Meanwhile, there was a change taking place in the English-speaking world. Papermaking had come very late to England. It was tried briefly but unsuccessfully in the early 1490s. The first English paper appears in Wynkyn de Worde's Bartolomaeus Anglicus: *De proprietatibus rerum,* of 1495 or thereabouts. This is documented by a statement to that effect in the book's colophon that it is the first English paper. But English papermaking doesn't really begin until 1588 when the first continuing mill in England was founded.

Although papermaking comes to England relatively late, the process whereby it came to England and stayed in England is quite interesting. You may know that the Edict of Nantes, which was published in France at the end of the 16th century, permitted toleration of Protestants. In 1685, the Edict of Nantes was revoked and Protestants, many of them papermakers (there were many Huguenot papermakers) left France, went to England and northern Holland, both Protestant countries. In both cases, they reinforced the papermaking industry, and those in Holland started the Dutch paper industry on a large scale, which was to make the Dutch the masters of the world paper trade in the 1700s. A proclamation was published by James II in 1687 that prohibited the papermakers in England (who established the White Paper Company) from leaving England by being enticed back to France by Louis XIV.

Papermaking came to America with William Rittenhouse in 1690. Rittenhouse was a Dutch immigrant, whose name is thought to have been Rittinghuysen, but his exact origins are somewhat uncertain. The first American papermill was established on the banks of the Wissahickon Creek in Germantown, Pennsylvania, in 1690 by William Rittenhouse, his son Klaus, and a partner, William Bradford (not the governor of the Massachusetts Bay Colony). William Bradford later became New York's first printer and published the first newspaper in the colonies. Unfortunatley, there is very little that is documented about the establishment of the Rittenhouse Mill.

American needs for paper continued to grow, but were, until the Revolution, supplied by imported, mainly English, Dutch, and Genoese, with some Spanish paper used in the distant South. Lack of trained craftsmen and, later, tax laws imposed by Britain to maintain the American colonies as agrarian consumers of British manufacturers kept the native industry small. The blockade of the American Revolution, national pride, and the industrialization that followed changed all that.

Hollander Beater

I spoke before of the Hollander beater. We don't see the Hollander beater in print until 1718 when it is shown in a rather famous engineering book by a man named Leonhard Christoff Sturm, *Vollstaendige Muehlen Bau-Kunst,* which was published in Augsburg (*Fig. 7*). It's thought that the beater began being used in the 1670s or 1680s in Holland, hence its name.

In the Sturm book, you can notice that the roll (beater roll) is surrounded by bars, much as a modern beater roll is. In this case, however, it runs against a flat bed plate rather than one that also has bars in it, for that is a refinement that comes along later. But the principle of the beater—with the roll rotating in one direction, the backfall to impel by the force of gravity the pulp around the vat, and the mid-feather of the vat separating the side with the roll on it from the side on which the stock travels—are all part of the earliest development of the Hollander beater. The chests, of course, were wood; the bars were iron or bronze. But it is essentially, in conceptual form at least, the modern beater that was used almost exclusively in paper manufacturing in the U.S. until relatively few years ago. It now has been displaced, for the most part, by continuous refiners of a conical type, which we call either Jordans of Claflins, or flat disc refiners (Bauer), or whatever, depending on their particular design.

As can be seen in a ground plan of a mill from the same Sturm book, the beaters were used in series. There were six beaters, all driven off one line shaft, with that line shaft geared into another line shaft, which was driven by a water wheel. The Dutch used both water wheels and windmills to drive beaters, both of which have have enormous cogwheel teeth on the gearings. Anyone who has stood next to a Dutch water wheel has some idea of how big those teeth are. Because reduction gears were practically unknown, it took an enormous amount of

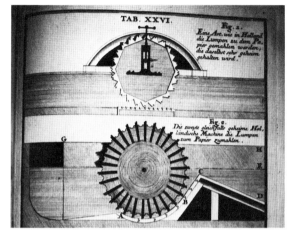

7. Leonhard Christoff Sturm. Vollstaendige Muehlen Bau-Kunst. *Augsburg. 1718. Entire page: 13" x 8". (Upper half only): Earliest illustration of the Hollander beater, which had come into use almost 50 years earlier.*

power to drive these wheels, which were virtually direct drive.

The Dutch were great at publishing mill books, too. Their mill books were advanced blueprints, much more carefully done than the machinery books of a hundred or more years earlier. These books contained a complete enough drawing of a beater and a set of views all drawn to scale (with the scale illustrated at the top), to make it possible for a millwright to take this sheet and build a beater from it, using it as plans. To give you an idea of scale, one book was 26 inches high and the plate some 22" × 36". It's pretty difficult to conceive of printing an engraved plate at that time in that size, but they were commonly done, in the Dutch mill books, in particular.

References in Early Encyclopedias

There were references in early encyclopedias to paper and printing. One of the ones on printing that's often lost sight of, and I commend it to you highly because this book can be found in libraries, is John Harris's *Lexicon Technicum,* published in

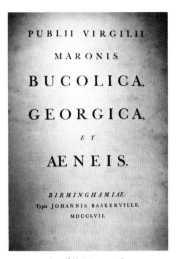

8. *P. Virgilii Maronis.*
Bucolica, Georgica, et
Aeneis. *Birmingham. John
Baskerville. 1757. 11 3/8" x
8¾". Title page of the first
book printed on wove
paper.*

1704 (first volume) and 1710 (second). It is the first technical encyclopedia in *any* language and the first encyclopedia in which there is a lengthy description of printing. In point of fact, it is the second printing "manual," the first being Moxon's *Mechanick Exercises* of 1688.

The wooden press shown in Harris is really almost the final design of the wooden press. It was used in virtually all of the printing shops in Europe and may still be seen in the exact form in which it was used in the Plantin-Moretus Museum in Antwerp, which gives any visitor the best possible depiction of a European printing shop of the 17th century. The needs of the wooden press in terms of paper were pretty limited, so surface refinement of paper is not part of the typographic printing process, as it is part of the engraving or intaglio printing process, and really doesn't change until the invention of the first iron hand press, and then the cylinder press.

Illustrational Treatises

Illustrational treatises are beginning to appear in the 17th century. The first and greatest treatise on engraving was Abraham Bosse's *Traité* of 1645, which was sufficiently popular to be reprinted three times within the hundred years following its first printing in 1645. The author gives not only a complete description of how to make engraved plates but a complete description of how to print them. He also has very distinct qualifications that he attaches to paper to be used in the engraving process, mostly having to do with surface. But it's not until 1761 that the first compendium on papermaking really appears.

The Académie Royale des Sciences in Paris conceived of a great project, an encyclopedia of *Arts et Metiers* to be published by them in fascicles. The first fascicle of 1761 published an article by Jerome Joseph de Lalande called *Art de faire le papier*. It contains a lengthy text of 160 pages, which described the papermaking process and all the laws attached to it. It contains 14 plates illustrating the whole process, and it was translated into German the next year. Lalande, contrary to popular opinion, is not a manual as much as it is an encyclopedic survey of the practices of the time. Despite its encyclopedic nature, it

is in fact the first complete published account of papermaking practices, including the couching operation.

In the German edition, the plates were recut because it's a quarto rather than a large folio, and it appears in 1762 under the name of *Schauplatz der Kuenste und Handwerke,* compiled by Von Justi. It is the identical encyclopedia of *Arts et Metiers* translated into German from the French. We see a rather defined form of stamper because stampers were still in use a hundred years after the invention of the Hollander beater. Here again, it is drawn fairly to scale in technological form to illustrate to the reader what the practices of the time were.

Lalande was also translated into Spanish in 1778 and, like the German one, it is rare since there are only two known copies of the Spanish translation. Smaller in size again, and also with the plates recut, Lalande finally appears in Dutch in 1792, translated by a Dutchman named Kastelijn. Lalande was finally translated in abominable form a year ago into English, although an earlier English translation had appeared in an abbreviated version in the *Gentleman's Magazine* in 1762.

We see in Lalande's book the process that gave rise to the first occupational disease in the paper industry, from the sorting of rags. It was well known that people working in rag rooms suffered from lung troubles as a result of inhaling the very fine fibers that were flying around in the atmosphere from the tearing up of rags. Papermaking in the 18th century was really rather a dirty occupation. You can imagine what the sorting and tearing of rags was like because clothing was worn to death in those days. Paper mills bought the used clothing from rag collectors whose rags were filled with all sorts of filth and vermin, and Lord knows what else. In the mills, they were taken apart by women in the rag room, dropped down through the floor into tanks of water, where they were washed and left on the side to ret. The entire papermaking process was a wet, sloppy one.

Wove Paper

The development of wove paper is the next landmark in papermaking history and is enormously important. The inventor of wove paper is fairly clearly documented. John Baskerville, the English printer who had made his fortune in the japanning business and who late in life went into the printing business, did not like what he perceived to be the low quality of English printing at the time and made up his mind that he wanted to do a better job.

In 1757 he published his first book, his great *Virgil* (*Fig. 8*), and the *Virgil* is printed on the first wove paper known to have been used in a book and certainly among the first wove paper ever made. It was made by James Whatman for Baskerville and is curious because you can still see chain lines in this paper even though it is wove. The reason is pretty clearly that Whatman probably took woven wire cloth and simply covered a laid mould with it, so the chain lines are visible because of the woven wire resting on the ribs of the mould. It is, nevertheless, wove paper.

Strangely enough, Baskerville published 30-odd books, but though he liked wove paper for a reason I'll describe in a moment, he used wove paper in only three of his books, in the *Virgil* of 1757, in the second volume, *Paradise Regained,* of the second issue of his Milton of 1759, and finally in his *Aesop's Fables* of 1763. The rest of Baskerville's books all use laid paper.

The reason Baskerville wanted to use wove paper is because of its surface, as I have been emphasizing all along. Baskerville's types had very fine serifs and major contrast between thick and thin strokes. In order to print those types satisfactorily, he felt he needed the refined surface that wove paper gave him. Further, he did not believe in having a lot of impression in the paper, so that even after printing on this relatively smooth paper, he hot-pressed the printed pages between copper plates to take the impression out and to polish the surface of the paper. The *Virgil* is a beautiful book by any standard. It is not particularly rare, but it is very important in the history of papermaking.

One of the places Baskerville's invention is documented is in a little book of 1786 about five inches tall, of Didot L'aîné, in which the "Fables" are followed by an epistle on the progress of printing, "Epître sur le progrès de l'imprimerie." In a lengthy footnote to the latter poem, Didot makes the point that it was Baskerville who invented wove paper, which came to France relatively late, and that it was clearly not a French invention.

At the time, it was being used by three French papermakers, including Johannot and Reveillon, who were disputing who invented wove paper in France. Didot's footnote was an attempt to resolve the dispute.

Paper Without Rags

In 1742, *L'Histoire des insectes,* a book by the great French nationalist Réaumur, was published. It contained an article that he had published earlier, in 1720, which had the following conjecture in it. He had noticed that wasps' nests seemed to be made out of a paperlike substance. Reaumur hypothesized that if wasps, which he had observed chewing on rotten fenceposts, were able to go from the fenceposts to making the nest out of a paperlike material, was it not feasible for man, somehow, to chew up wood and make paper out of it. This was speculation on his part and really led to nothing more than its being picked up by several other people. But experiments began in the later part of the 18th century to find raw materials other than the cotton or linen rags or hemp that had up to that time been the sole source of papermaking materials in Europe.

The same wasps' nest conjecture appeared in a book by Jacob Christian Schaeffer, a German naturalist and preacher of Regensburg in 1765, when he published a great work in the history of papermaking and a great landmark, called "Research and Samples to make paper without any rag or with simply an admixture of rags" (*Fig. 9*). The book contains (in its six volumes published between 1765 and 1772) some 86 different specimens of paper made out of a variety of different materials by Schaeffer, who attempted to see what could be done with materials other than cotton, linen, or hemp. The second sample in the first volume is wasp-nest paper.

One substance Schaeffer made paper out of was *Tannenzapfen* (pine cones), which results in paper that is brown-colored, the natural color of the pine cone. Also, he painted the paper or had his daughter paint it to demonstrate that it could be used as a writing or painting material.

There's an amusing sequel to Schaeffer's book, however. In the 1920s, Wilhelm Herzberg, the great German paper chemist and the father of laboratory paper testing, published

9. *J. C. Schaeffer.* Versuche und Muster . . . *Regensburg. 1765. 7½" x 5½". First German edition. Title page of Volume I of Schaeffer's 6 volume work on experimental papers.*

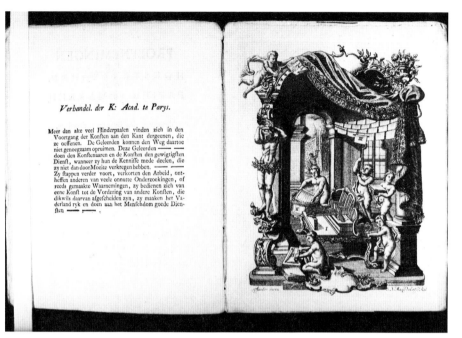

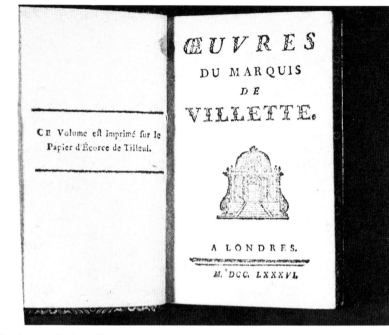

10. J. C. Schaeffer. Proefnemingen en monster-bladen om papier te maaken. Amsterdam. 1770. Fanciful illustration of stamper used by Schaeffer for preparing pulp.

11. Charles Marquis de Villette. Oeuvres. Lordres (Paris). 1786. 4 5/8" x 2 7/8". Title page spread of the first book printed on paper made entirely without rag. The facing page describes the paper as having been made from "live-tree" (linden) bark.

a little book *Schaeffer's Researches Into Paper* in which Herzberg, in very Teutonic manner, took slivers out of all 86 specimens of Schaeffer's book and found, I'm sorry to say, that Schaeffer cheated. They all contain some rag fiber, probably in order to hold the sheets together, and some samples are mainly rag fiber. Schaeffer was a great pioneer, nevertheless. Herzberg's work is deadly serious and amusing in our eyes because of that, but that doesn't in any way reduce the importance of Schaeffer's researches. Schaeffer used a little stamper, carpentered in an ornate manner, to prepare the samples illustrated in his book.

The first great researches into making paper without rag at all were made by a man named Pierre Léorier de Lisle, who was the manager of the royal papermaking factory at Montargis near Paris. He published (in 1784) a few specimens in a very rare little book, *Les Loisirs des bords du Loing,* which is about five inches tall. One is a sample of green paper, made of grass, and in a very French way, it contains a poem appropriate to the paper.

Two years later, De Lisle published a little book (*Fig. 11*), the *Oeuvres* of the Marquis de Villette, which appears in two forms, one on paper made out of linden bark (*ecorce de tilleul*) and the other version on paper made out of "guimauve," which is a reed called the marsh mallow, not the sugary confection we think of, but a reed like a cattail. The book contains 20 rather beautiful specimens of De Lisle's experiments, all of them paper made out of vegetable substances containing no rag, including one with leaves of burdock and a weed called sheepsfoot. The samples are lovely, and the book is beautiful in every way, although it is pretty rare.

12. Paul Sandby. Views in South Wales. *1775. Detail approximately 1¾ " x 2¾ " from "Benton Castle." Plate XI. Aquatint.*

The first reference to papermaking materials in America was published in Philadelphia in 1777. It contains a reprint of an English book, *Select Essays,* originally published in 1754, which contained the researches of a man named Guettard, who may well have been inspired by Réaumur into trying different materials that might be used for making paper. Guettard was a physician at the court in Paris and did his own little experimentation with natural materials, which was not really very successful, but his researches were published.

Aquatint, Wood Engraving, and Lithography

I spoke before of illustrational processes. One of the things that was taking place in the 18th century was the attempt to develop ways of producing illustrations that would better reproduce pencil or watercolor drawings, their washes, and all their subtleties.

The first attempt along that line was the crayon method of etching, a mechanical method which was done with a roulette. The method was invented by Jean Baptiste Francois in Paris and, beginning in 1759, he published nine volumes of portraits of "Philosophes" of all nations.

At almost the same time, the process of aquatint was beginning to be developed and, for the first time, it was possible to reproduce watercolor washes almost as they appear in paintings. Aquatint was invented in France by J. B. LePrince and eventually found great popularity in England because it's peculiarly adaptable to English watercolor technique and the English taste for landscape. The first published English aquatints were Paul Sandby's *Views in South Wales* of 1775 (*Fig. 12*), which call upon paper to do some things that it had never

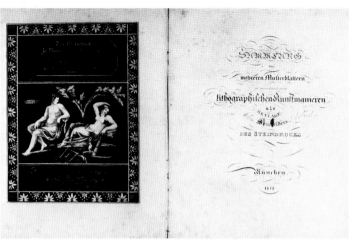

13. *Thomas Bewick.* "The Departure" *from* Poems *by Goldsmith and Parnell. London. 1795. Detail approximately 1¾ " x 2¾ ".*

14. *Alois Senefelder.* Vollstaendiges Lehrbuck der Steindruckerey. *Munich. 1818. 12 1/8 " x 8 7/8 ". Title page and one plate from a volume of the inventor's own manual of lithography.*

15. *Johann Nepomuk Strixner.* Albrecht Dürer's Christlichmytholorische Handzeichnungen. *Munich. 1808. Image area 11 1/8" x 8". The earliest book using lithography as a reproductive medium.*

been called upon to do before. Its very fine lines and washes were extremely difficult to print on anything except paper that was selected, if not made, specifically for the purpose.

The process that was to take over the whole 19th century, however, was wood engraving. Thomas and John Bewick were really the people responsible for the revival, if not the invention, of wood engraving (*Fig. 13*). Wove paper played an important part in these processes because wood engraving almost never appears on anything besides wove paper since the fine line of white on black technique cannot be reproduced nearly so easily on laid paper.

In the case of lithography, Senefelder points out in his manual of lithography (1818) the fact made in every lithography manual of early times that wove paper should be used in lithography, too. Senefelder was as good as his word because, although the text matter of his book is printed by letter press on laid paper, the plates are printed on wove paper (*Fig. 14*). Senefelder also experimented with color lithography. The green is painted by hand, but the silver, blue, and black of a

plate from his 1818 manual are printed in color. It's color lithography earlier than most people give it credit for.

The first artist's lithographs (and the reason I dwell on lithography is because lithography might not have been possible had it not been preceded by the invention of wove paper) appear in England in 1803 in *Specimens of Polyautography* which contains, among others, a famous plate by Benjamin West. Most of the prints are on wove paper, and the paper has not browned (it's tinted paper in most cases). When we inspect a closeup of West's lithographic technique, it may be seen that the attempt on the part of West and his cohorts to imitate pencil drawings was very successful indeed, in no small part because the required materials were available at the time.

Reproductive lithography begins in Germany at the same time, also on wove paper. Johann Strixner's *The Prayer book of the Emperor Maximilian,* illustrated by Albrecht Dürer, exists only in manuscript form (*Fig. 15*). The book was drawn from the manuscript by Strixner in 1808 and was published in lithographic facsimile. A few years later, in 1819, Strixner did early two-color lithography, in the making of his facsimile of the Cranach portion of the same prayer book. It's printed mainly in green, except for the church at the bottom, which is printed in red in register, possibly from the same plate by inking it through a stencil.

Confluence of 1800

You may wonder why, except for the element of wove paper, I am discussing lithography in a talk about paper history. The reason for doing so is that lithography, or *Steindruck* (stone printing, as it was called at first), was one of the elements in what I have chosen to call the "Confluence of 1800," a series of events that took place *about* 1800, all involving the graphic arts and all exerting profound influence on revolutionizing communications.

The invention of lithography is datable, and although early experiments began in 1796, it was perfected in 1798, as Senefelder himself documents, and began to be used about 1800. One of the other elements of the confluence of 1800 was the invention of the iron press. Earlier, I spoke of the "Blaeu" press, which was made of wood and which in modified form was still in use until 1800. In 1800 the Earl Stanhope, in England, invented the iron hand press, called the Stanhope press. It was followed closely by the Columbian press, which was invented by an American named George Clymer, and that in turn was followed by the most common of all iron hand-presses, the Hoe Press, also called the Washington Hand Press.

Iron presses displaced wooden presses, except for the backwoods areas, within relatively few years. This invention, along with the development of lithography and the third element of the confluence of 1800 — the invention of the paper machine — caused radical change in the graphic arts.

The first paper machine was invented by a Frenchman named Nicholas Louis Robert, an employee in the paper mill owned by the Didot family and operated by Leger Didot in Annonay. One may yet see Robert's own drawings (*Fig. 16*), which were drawn by the inventor to send to England in the hands of his brother-in-law, John Gamble, to apply for the British patent for the papermaking machine, a patent which

16. *Working drawing of longitudinal section of the first paper machine, prepared for the English patent application by Nicholas Louis Robert, the machine's inventor. 18¼ " x 24 7/8".*

was granted in 1801. The patent development was financed and backed by Henry and Sealy Fourdrinier (an English name, though French-Huguenot in origin). They went broke in doing it, but in going broke, they gave their name to the papermaking machine, the Fourdrinier paper machine. On my Robert drawing, both John Gamble's name and Henry Fourdrinier's name or that of Bloxham and Fourdrinier, the firm, are there in the lower left.

The machine operated by turning a hand crank at the end of a geared shaft, and the shaft, in turn, moved the wire belt in

17. Matthias Koops. Historical Account. *London. 1801. 11" x 6¾". 2nd edition. Illustration engraved on paper made of straw. Title page printed on paper made from waste paper.*

Matthias Koops was the first man to make paper commercially out of straw, waste paper, and wood pulp. Koops's book is considered to be the third of the "Bach, Beethoven, and Brahms" of papermaking—Schaeffer's and De Lisle's being the other two. One illustration in his second edition (1801) is on a yellowish paper, which is actually straw paper (*Fig. 17*), while the book itself is on paper made from waste paper (recycled paper) and also out of wood pulp. The first edition of his book, *Historical Account,* was published in 1800, the year he started his mill in London, called the Neckinger mill. Unfortunately, the mill closed down four years later when he went bankrupt.

One of the causes of the deterioration of paper is thought to be the excessive use of alum in sizing paper, which gives it too much acidity. The invention of sizing paper in pulp form, rather than in finished sheets took place in about 1800.

Moritz Illig, the inventor of rosin sizing, printed and sold (with the rights to his process) a handbook, which is extremely rare with only two copies known to still exist. Illig sold the book for 500 gold guilders, which entitled the buyer to the rights to the process. The point to be made here is that the title of the book itself, *Anleitung...Papier in der Masse zu leimen,* tells that it refers to a way of sizing paper "in der Masse"—in pulp—rather than in sheets, as it had been done before.

a longitudinal direction. The paper was rolled up on the top of one of a pair of felt covered rolls, and the higher roll moved upward as the paper built up. The machine would be stopped, the roll of paper would be cut off with a knife, unrolled, and laid out to dry.

Interestingly enough, it is apparent that one of the reasons Robert invented the paper machine was to make paper in rolls long enough to make wallpaper, because wallpaper was enormously popular at the end of the 18th century. Until his invention, wallpaper had to be made by pasting sheets of handmade paper together on the wall. With the paper machine, it became possible to make wallpaper in the now-familiar short rolls and, for the first time, to put up floor-to-ceiling panels in a single piece. The paper machine was soon modified by a great engineer, Brian Donkin, who was responsible for the design and building of all of the early "Fourdrinier" paper machines. By the 1820s, the paper machine was in wide use in England.

Chemistry and Testing

Chemistry came into play in the late 18th century as bleaching with chemicals began to be used for the first time. Scheele had discovered chlorine in the 1770s, and bleaching of rags with chlorine became common by the 1790s. Straw paper, too, could be bleached. A book by Louis Piette, the great French paper chemist and papermaker of the mid-19th century, shows 250 samples of different papers made out of straw, some bleached, some unbleached. Without the chemistry that had preceded him, this would not have been possible.

The search for raw materials continued. In 1822, William Cobbett published *A Treatise on Cobbett's Corn,* which contains paper actually made out of corn husks. In it he ventures to suggest that since corn had recently been found in America, the husks could be used to make paper. In the 1866 *Practical Guide for the Manufacture of Paper and Boards,* Prouteaux

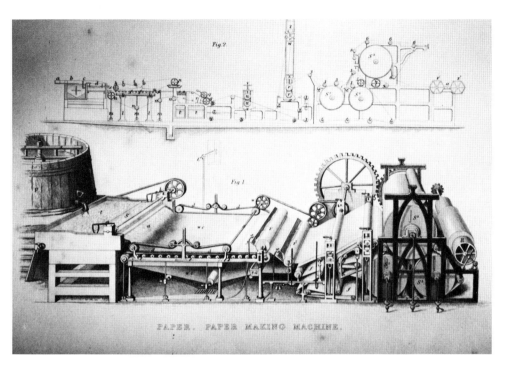

18. Paper machine of the mid-nineteenth century. From Tomlinson's Cyclopaedia. *London. 1855. 10 3/8" x 6¾". The second Industrial Revolution and the advent of steam power made changes in the paper machine. This is a paper machine of 1855, and really, except for a few frills, is like Robert's paper machine in principle and like the Fourdrinier paper machine of today.*

discusses the advent of wood pulp in American papermaking for the first time. The American Wood Paper Company was established in Manyunk, Pennsylvania, near the present mill of Weyerhaeuser Corporation, formerly W. C. Hamilton Co., on the Schuylkill River. On that river the American Wood Paper Company made the first commercial chemical wood pulp in the United States.

William Savage's *Practical Hints on Decorative Printing* is a great book for its illustrations. One plate is printed from 23 woodblocks, printed in register. It's a real tour de force, as is the rest of the book. Savage was trying to demonstrate that relief printing, as late as 1822, still had an important place in use. Again, he required wove paper to do his printing.

Savage's book has added importance for our purposes today because it contains one of the earliest articles on paper testing. The article was written by Michael Faraday, who tested and analyzed India paper, then also called China paper—

paper that was heavily filled with clay, which was made in the East and imported into England.

China paper was notably successful for the printing of illustrations. Savage knew that by experience, so he asked Faraday to analyze the paper and find out why. Faraday found out that the filler in the paper (which he found by burning it and weighing the ash that was left over) had a direct influence upon the surface of the paper. The filler closed up the paper, making it possible to print illustrative material more readily. It was the beginning of a whole chain of events that culminated in coated paper.

Colored paper was used fairly early in the 19th century. The history of colored paper is interesting because the earliest colored paper that has survived was made in the early 16th century, about 1515-20, and it is always blue paper. Aside from the popular stories about a bag of laundry bluing falling into the papermaking vat, we don't really know how blue paper came to be used, but we know what it was made of. Blue paper prior to the mid-19th century, with the exception of a few papers made with natural dyestuffs, was made either by using colored rags or by using mineral pigments to put in with the pulp in order to color it. The only dyes (as distinguished from pigments) that were available at the time were natural dyes, primarily indigo for blue, cochineal and logwood for red, and some other natural dyes that made various yellows. The dyeing industry, as it existed until the second quarter of the 19th century, was devoted to textiles, and paper coloring was given no thought. We begin to see color *printing* in the 1840s. Thomas Shotter Boys's *Paris, Ghent and Rouen* (published 1839) is the first book executed in chromolithography. Here, again the use of a refined paper surface is required because the plate is not inked in portions and printed in one impression. Instead, there are several impressions of color printing in register that require the use of anywhere from four to 12 lithographic stones. The refined surface demanded wove paper, the necessary predecessor to lithography and in this case, color lithography.

Later in the century, a reproductive revolution occurred that started with the invention of photography (and the eventual development of photomechanical processes) and culminated in the creation of the halftone screen in the 1880s. These great advances could not have taken place without refined paper surface accomplished, first, by smoothing paper, second, by using fillers to close up the interstices between the fibers and, finally, by developing coated paper, formed by adhering clay coating to the surface of paper with casein adhesive.

Good halftone printing by the relief process is unobtainable without a coated surface. In offset lithography, the coated surface enhances the appearance of inks on the surface and is commonly used, although it is not necessary for reproduction.

Dard Hunter's Legacy

I would like, finally, to speak of Dard Hunter, who is really responsible for the rediscovery of and the recent revival of hand papermaking in the United States. Dard Hunter worked as a designer for Elbert Hubbard at the Roycrofters Press in East Aurora, New York, long before he wrote anything about papermaking history. Hunter went on from there, however, to begin to write and publish books about papermaking, for many of which he cast the type, made the paper, and printed the text himself.

His early books exemplify his influence, which I consider to be tripartite. In the first place, Hunter brought primitive papers to the attention of a small, but nevertheless significant, worldwide public. There are samples of tapa in Hunter's book *Prmitive Papermaking,* published in the 1920s, that illustrate his interest in primitive paper. Second, Hunter was interested in Oriental papers. His great work *Chinese Ceremonial Paper,* published in 1935, is probably the most difficult to find of all Hunter's books because it contains the largest number of samples, all of them unobtainable today. The value of Hunter's books lies in the samples that they contain because, just as the samples of these Chinese papers simply could not be assembled again, neither could many of the samples in Hunter's other books.

Hunter's third influence is in the field in which he did do a primarily scholarly service, his researches into the origins of American papermaking in his final book, *Papermaking by Hand in America,* published in 1950. Hunter described many of the early American paper mills, and the first mill in each state is documented here, as well as many of their watermarks and some of the ream labels that they used. Hunter's influence cannot be underestimated.

Revival of Hand Papermaking

John Mason, an English teacher of bookbinding at Leicester University, has also had a great influence upon the revival of hand papermaking. A small book of his, *Papermaking as an Artistic Craft,* 1962, describes papermaking with various materials in a kind of Schaefferesque way with a delightful sense of humor.

Another pioneer is Henry Morris, of the Bird and Bull Press, who wrote *Omnibus* in 1967, which tells of his experiences in hand papermaking. I would commend it to you, as I would the Mason book, to anyone who is seriously interested in making paper by hand. Bear in mind, please, that all the paper

I've talked about (and Morris's paper and Mason's paper are no exceptions) is paper to be used as a substrate rather than an art form of itself. These writers don't go into the making of that kind of paper, except that Mason touches the edges of it with the use of leaves and vegetables and other objects contained or embedded in paper.

Finally, I would like to talk briefly about paper as an art form, which has also been revived. Decorated paper is a whole field by itself, and one on which I could spend an equivalently long time as I have on this. Suffice it to say that decorated paper has always been of interest in the West and really does not, at least in the form of marbled paper, come into Western use until after 1600. Marbled paper was invented in Turkey. To gain the kind of marbled effect that its makers sought, the surface was refined by glazing the paper with a polishing stone. In Turkey, "silhouette paper" was made by printing (it is thought with leather) colored inks on the surface of refined paper, and in some cases, the surface of paper was colored by putting gypsum and some pigment on the surface actually as a coating. That's called "aher" paper.

Obviously, my purpose has not been to deal with history per se, nor is it to turn the clock back. My presentation has simply been an attempt to put into perspective a portion of the history of papermaking. I've not discussed recent technology, nor do I have time to do so. Paper has been made out of wood pulp since the 1860s in increasing volume. Sizing materials and other additives have been put into paper since the early 19th century. We in the paper industry have learned a great deal about the making of paper in the last 20 years, having to do principally with the improvement of the longevity of paper.

I would like, in conclusion, to say that it is not appropriate to look only at the materials from which paper is made in order to be critical of it. We must look at the way in which it is made and at the way the paper appears in final form. I think that it is a two-edged sword. I think that materials used for paper-making may be of themselves advantageously used but may be impermanent. For any of you who are making paper for art works, you may wish to strive for permanence. If you do, I wouldn't worry as much about the specific materials you use if you use them carefully and as long as you use them in relatively pure and uncontaminated form and try to stay in the areas about which we have learned something from chemistry.

Leonard B. Schlosser, involved with the paper industry since 1947, has taught at New York University and is a member of numerous rare book organizations and the International Society of Paper Historians. His collection of books and documents related to the history of paper and writing material is widely known and includes all the pieces used in this article. He is president of the Lindenmeyr Paper Corporation in Long Island City, New York.

JAPANESE PAPERMAKING: SEKISHŪ-HANSHI

Yasuichi Kubota

Translation by Hitoshi Sasaki

Introduction and annotation by Timothy Barrett

1. Akira Kubota, Yasuichi Kubota, and Timothy Barrett at the International Paper Conference, in San Francisco, March 1978.

Introduction

Sekishū literally means "rock country" or "rock state" and is the old name for the western area of Shimane prefecture in Japan where this famous paper continues to be made. *Han* means "half" and *shi,* "paper." Thus, a rough translation of *Sekishū-hanshi* is "rock country half paper". The term "half" refers to the way the finished sheets were cut down from larger sheets. The term has come to refer to a traditionally made *kōzo* paper from the Sekishū or Iwami area.

Hundreds of years ago, *Sekishū-hanshi* was used chiefly as pages in the account books of merchants in Osaka. The books were reputedly thrown into the nearest well if fire broke out, only to be recovered, dried, and put back into service after the calamity had ended—traditional testimony to the paper's strength and reliability.

According to regulations issued by the Emperor Nombu in 701 and the Emperor Daigo in 905, *Sekishū-hanshi* was offered to the Imperial Court at those times. Thus, its origin can be dated well before A.D. 1000. Evidence of local contact with China, however, suggests an even earlier date for the first *Sekishū-hanshi*.

A guild of houses producing *Sekishū-hanshi* was organized in 1904, contributing greatly to production and distribution of the paper. Since the Second World War, however, social and industrial changes have resulted in a drastic decline in houses making the paper and raising the special fiber. In 1969 the Japanese government designated the craft of making *Sekishū-hanshi* an "Intangible Cultural Property of Importance," thereby calling attention to its existence and joining in the cause to stave off further decreases in its production.

Yasuichi Kubota was born in 1924 and grew up in a papermaking household (*Fig. 1*). Today, he produces *Sekishū-hanshi* on a year-round basis and heads the *Sekishū-hanshi* Technical Association. His papers include not only the traditional *kōzo* paper but also *mitsumata* and *gampi* paper and an extensive range of natural vegetable-dyed papers. Kubota is one of only seven craftsmen still producing *Sekishū-hanshi*. His son Akira, aged 27, is the only young person in the area who has made a commitment to continue the craft. Akira accompanied his father to the 1978 conference "Paper — Art & Technology" in San Francisco to assist in the demonstration.

(Editor's note: The following is a synthesis of Kubota's speech and comments on the craft made by Timothy Barrett. Barrett's annotations are indented and are meant to clarify Kubota's process as well as place his work in contrast to the mainstream of modern papermaking in Japan.)

History of Japanese Paper

Papermaking began in Japan about A.D. 610, introduced through China and Korea. Throughout the past 14 centuries, both the methods and materials of Japanese papermaking have been researched and improved, and three plants have proven to be the most suitable.

> The bast fiber used in Japanese papermaking is produced from the white inner bark of young *kōzo* (*Broussonetia kazinoki,* Sieb.), *mitsumata* (*Edgeworthia chrysantha,* Lindle), and *gampi* (*Diplomorpha shikokiana,* Honda) trees. All three fibers are extremely long even after beating, ranging in length from 3 to 12 millimeters. By contrast, the lengths of beaten Western papermaking fibers are 4 millimeters and shorter.

The longevity of papers made with these plant fibers is demonstrated by documents stored in Shōsō-in, in Nara City, Japan, where after 1,300 years many of the papers remain as beautiful as freshly made sheets. Many of the papers made in the last one hundred years have not survived as well as these traditionally made papers. The reason for this becomes clear when one analyzes the methods and ingredients that have been introduced since the early Meiji period.

> The Meiji period (1868-1912) that Kubota mentions here marked the beginning of intense Western influence in Japanese history. A variety of machines, chemicals, and wood pulps were gradually incorporated into the traditional craft that have since proved detrimental to the character, permanence, and durability of the paper. Kubota is one of only a few paper-

makers in Japan today who uses only bast fiber and highly traditional processes during its preparation.

Cultivating and Processing the Trees

Kōzo and *mitsumata* can be grown in cold, or even snowy areas, but both prefer mild climates with average rainfall. *Kōzo,* a short deciduous perennial and member of the mulberry family, can be harvested one year after planting but usually is not harvested till the second year when the yield is greater. New sprouts appear each spring from the same root and are cut yearly. After five to eight years, the harvest amount from one root is optimum. *Mitsumata,* a deciduous perennial of the Ginchoge family, is the newest Japanese papermaking fiber and was first used in the late 16th century. Both *kōzo* and *mitsumata* are harvested in the winter, before buds begin to sprout and when the bark is at its thickest.

Steaming

After harvesting, *kōzo* trees are cut into one-meter lengths and steamed in a process called *seiromushi,* for at least two hours, until the skin can be pierced with the fingernail (*Fig. 2*). Then the bark is stripped off and hung to dry (*Fig. 3*). Later, it is soaked in water and scraped with knives to remove the black outer layer and usually the green middle layer, leaving only the white inner bark (*shirokawa*) for papermaking.

> In a few areas (Kubota's village of Misumi being one of the most famous), the green middle layer of bark is purposely left attached to the white bark at this stage. Papers made from both the green and white bark are not as light in color as those made from the white bark alone, but according to Kubota and others, the resultant paper is stronger, crisper, and more resistant to the ravages of insects. These effects are due to the presence of large amounts of various hemicelluloses between the green and white layers which contribute to strength and crispness, and to essential oils present in the green layer which may help ward off insects.
>
> The "Ginchoge" tree Kubota mentions is *Daphne odora* Thumb., of the order *Thymelaeaceae.*

2. Kōzo *trees, bundled in one meter lengths, are steamed in a process called* seiromushi *to soften the bark for stripping.*

3. Kōzo *bark being stripped.*

Gampi

Gampi is also a short, deciduous, perennial bush belonging to the Ginchoge family. The trunk of *gampi* is smooth and dark-brown colored, and the full tree grows to about three meters. *Gampi* cannot be cultivated so it must be harvested where it grows wild in the mountains. *Gampi* grows best in warm sunny areas. The best fiber comes from the Shizuoka, Kinki, and Shikoku areas. Five-year-old *gampi* trees produce the best fiber and must be cut in the spring and stripped just after cutting. In this way, *gampi* does not need to be steamed like *kōzo* and *mitsumata.*

For papermaking, generally, it is better to use *gampi kurokawa* (black bark) as is, rather than making *shirokawa* (white inner bark) first. Paper made from *gampi kurokawa* shows the special gloss of *gampi.* Finished *gampi* papers are used for bookkeeping, gold and silver leaf production, special waxed reproduction papers, and for any use requiring long-term storage or resistance to insects.

Once stripped, as black bark and later after cleaning, as

white bark, *kōzo, mitsumata,* and *gampi* are hung up to dry and then stored (*Fig. 4*). The remaining inner wood of the tree is burned for fuel.

Today, *kōzo* paper represents more than half of the hand-made paper produced in Japan, and it could be said that *kōzo* paper is the most typical Japanese paper. It is not suitable for machine papermaking because of its extremely long fibers.

Boiling

The most important stage in papermaking is the preparation of the materials during cooking and beating. These Japanese plant fibers are fairly stable and resistant to alkalinity. Therefore, the fibers can be soaked and boiled in alkaline solutions without damage. Chemicals used for this boiling are woodash, limestone, soda ash, and caustic soda. The different kinds of chemicals used in the boiling process will produce different characteristics in the finished paper, so they are chosen carefully with the desired product always in mind.

4. Kōzo shirokawa *(white bark) drying in the sun.*
The bark is stored for future papermaking.

5. *The* shirokawa *is cooked in a chemical solution to dissolve undesirable constituents. A thorough rinsing follows.*

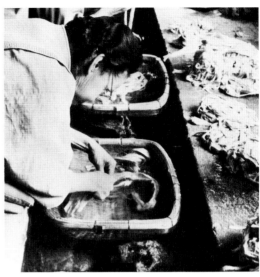

6. *Foreign particles are removed by hand from the fibers.*

The bark is cooked in a solution consisting of a chemical (added at 12% to 20% of the weight of the dry bark) and water (at 15 liters for each kilogram of dry bark). Cooking (*Fig. 5*), which lasts roughly two to three hours depending on the kind of paper being made, dissolves the nonfibrous constituents (waxes, gums, lignin, and pectin) in the bark. Surprisingly, these unwanted constituents account for half the weight of the dry bark.

After cooking, the fiber is put into a concrete tank and rinsed thoroughly with a continuous flow of water until all vestiges of the chemicals are removed.

Removing Foreign Particles

Foreign particles can be anything remaining in the cooked fiber as a result of damage to the live tree caused by frost, insects, or hail. In addition, traces of buds or unwanted specks of black outer bark may also remain. Although this is a very time-consuming, laborious process, all traces of these foreign particles must be picked out in order to insure beautiful paper (*Fig. 6*). This must be done by hand, although a modern cleaning device is sometimes used to clean *mitsumata* and *gampi*. This method, however, can cause damage to the fibers, thereby producing inferior paper.

The device to which Kubota refers is a "flat diaphragm screen," a machine with slotted plates somewhat similar to a European knotter. Kubota's fiber is always picked clean by hand.

Beating

This process is as important as cooking since it separates the fibers (*Fig. 7*).

Hollander beaters, as well as stampers and naginata beaters, are used by Japanese papermakers. The naginata is a Japanese device that looks very much like a Hollander beater except that it has no bedplate or backfall and, instead of being fitted with a roll, it has a set of curved knives mounted on its shaft. In action, the blades hack at the bast fiber suspended in water, sending it moving around the tub at a quick pace.

Regardless of the method, the purpose of beating in Japanese papermaking is only to loosen the fiber. The longer

7. *Beating separates the fibers and is carefully controlled, as overbeating will break and shorten the fibers and reduce the rate of drainage through the* su.

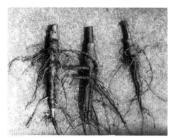

8. Tororo-aoi *root (*Hibiscus manihot, Medikus*), a major source of* neri, *the viscous formation aid used in Japanese papermaking.*

any beating continues, the shorter the fiber becomes while its freeness, or drainage on the mold, decreases. Beating is continued no longer than absolutely necessary because long fiber length and a high drainage rate are essential to successful sheetforming.

Tools for Sheet Forming

The vat, usually called a *fune* or *sukibune,* is made of pine, cedar, Japanese cypress, and more recently, concrete or plastic. Its size depends on the size of the paper being made, but it can measure up to four meters by four meters. The bamboo stick used for stirring is called *bashu.*

The *su,* or screen, is made of fine pieces of bamboo joined by silk thread. The *su* is flexible and removable. (Very similar to a bamboo placemat. — T.B.) This is in contrast to its Chinese ancestor which was fixed to the mold. The *su* is placed in a lightweight, hinged frame (mold and deckle) made of Japanese cypress (*Chamaecyparis obtusa* Endl.).

Neri

Besides the flexible *su,* another element discovered and developed by the Japanese papermakers is the addition of *neri* to the papermaking solution. *Neri* is a vegetable starch derived from the root of various plants, *tororo-aoi* (*Hibiscus manihot,* Medikus) (*Fig. 8*), being the most common. The root is crushed, yielding a liquid, called *neri,* which though viscous, is not sticky. The *neri* is filtered through a cloth bag to remove impurities.

The *neri* does the following kinds of work. It: 1. increases strength of the paper; 2. eases forming of very thin sheets; 3. increases hardness of the paper; 4. makes wet paper easier to peel apart; 5. prevents fiber from settling on bottom of vat; and 6. slows flow of liquid through the *su.*

Neri is not added as a sizing or adhesive agent, but rather as a formation aid, to disperse the long bast fiber and slow drainage of the vat mixture through the *su.* Not much *neri* is needed to accomplish its various functions. If the vat solution becomes too viscous, beating it with the bamboo stick will thin the fluid.

Forming Sheets of Paper

At the vat, the Japanese papermaker operates the mold with a distinctive sloshing-rocking motion (*Fig. 9*). He slides the viscous vat mixture repeatedly back and forth across the mold surface in the course of forming one sheet, picking up additional charges as required. As he does this, water drains through the mold surface, and he gradually layers or laminates the sheet to the appropriate thickness. This is a very different action from that of Western papermaking where the vat man takes one quick charge from the vat and shakes the mold only slightly, causing the fiber to fall into a sheet with a more felt-like formation.

After the papermaker has worked the sheet to the

9. The papermaker quickly scoops up the solution, shakes the mold from side to side to spread the fibers evenly, picking up additional charges that laminate together to create the desired thickness.

desired thickness, he stops, raises the deckle, and removes the *su* with the new fragile sheet attached. Very carefully then, he turns around and lowers the *su* like a curtain, with the paper side down, working it across a stack of previously couched sheets (*Fig. 10*). Once in place, he peels the *su* away, leaving the new sheet of paper smooth and unwrinkled on the pile. No felts are used between the sheets. The papermaker returns the *su* to the mold and continues making paper, accumulating perhaps 400 sheets in a day.

Wet paper is left to drain slowly overnight and then be pressed gradually the following day. The 18-pounds-per-square-inch pressure required in Japanese papermaking is slight compared to the 150 pounds per square inch exerted when Western handmade papers are pressed. Unbelievable though it is, the slightly damp paper can be peeled away from the stack one sheet at a time after pressing.

Drying

Even after pressing, the paper contains 60% to 80% water. There are two ways of drying: sunlight and artificial heat. Sunlight is, of course, traditional *itaboshi,* board drying. This is the typical image of Japanese papermaking (*Fig. 11*).

10. Having removed the su *from its rigid frame, Yasuichi Kubota lowers the* su, *paper side down, across a stack of previously couched sheets. He will then peel away the* su, *leaving the stack of* washi *to drain. No felts are used between the sheets of* washi. *(Photo courtesy of the Simpson Paper Co.)*

11. After draining, the finished sheets are brushed onto finely finished boards and left to dry in the sun. This is the traditional drying method still carried on by a few Japanese papermakers.

The material for the drying board has to be dense and fine wood and should be well dried and solid. Ginko, horse chestnut, pine, and Japanese cypress are commonly used. Brushes made of straw, hemp, horsemane, or horsetails are used to apply the paper to the board. For special paper, a leaf of camellia is used to give the proper sheen by rubbing the back side of the paper with the leaf. The side touching the board becomes the front side of the paper. Paper is dried artificially by being applied to a steel sheet, the other side of which is exposed to heat, hot water, or steam.

Sunlight drying has the following advantages: Traditional paper surface and touch will be obtained; paper is bleached by sunlight; it will not increase in weight after drying; and expansion and contraction rate is much smaller than paper dried artificially. Sunlight drying has the following disadvantages:

The paper is sometimes too soft; weather is often unreliable; color is dependent on changes in weather; and it is not suitable for mass production.

For its part, artificial drying has several advantages: The surface is tightened, thus strong paper is obtained; its surface is very smooth; it is very even and uniform; it is not dependent on the season; and it makes mass production possible. However, artificial drying has disadvantages: The special character of Japanese paper is lost; its weight increases with age; and it requires large amounts of fuel.

Outstanding Characteristics of Japanese Paper

Japanese paper is lightweight and porous and thus does not seal out the air. It is also soft and translucent, as well as strong and durable. Pure Japanese paper increases its gloss and hardness with age. It is used as stationery, for calligraphy and bookkeeping, as well as for drawing and painting. When Japanese paper is used for restoration of antiques, it is used not only for its beautiful appearance but also for purposes of permanence.

In Japan, there are many indispensable papers used in festivals or ceremonies, such as the lantern, *Toro tanabata-shi,* used at the Summer-Star Festival on July 7. There are also papers that require a great deal of durability, such as paper for gold leafing or paper for woodblock printing, which can be printed several hundred times. There are papers used in the process of printing kimono fabrics that are repeatedly treated with hot water, and papers used in making stencils for printing textiles where glue or starch are applied many times during the process. The uses of Japanese paper are endless.

When people judge the quality of Japanese paper, many of them judge only the shape or surface. When I look at most of the paper being made today, I wonder if it will stand up to the test of time. Already, paper less than one hundred years old, which has been made with the aid of modern machinery or the addition of wood pulp, shows great damage by harmful insects and decay. The life of traditional Japanese paper is its strength of resisting aging. This is the quality of Japanese paper that cannot be comprehended or understood by just momentary viewing. The true quality or beauty of Japanese paper will be finally recognized one or two hundred years after its creation. I am going to keep making the traditional Japanese white paper with delicate surface and touch and various physical varieties while attempting to return to the original heart of Japanese paper made before 1870.

Finally, I would like to thank Timothy Barrett, a researcher of Japanese paper, who studied in Japan for two years. He is a very good man, is enthusiastic, and has learned almost everything about Japanese papermaking. Shortly before he returned to America in 1977, he came to my house, and I remember my foolish words to him. I said, "If you make paper in Japan, you are capable of making *washi* (traditional Japanese handmade paper). But if you make paper in America, using exactly the same process, it would be very difficult for you to make *washi.*"

Timothy Barrett recently spent two years in Japan on a Fulbright fellowship studying the Japanese nagashizuki *papermaking technique, working part of the time with Kubota. He has just completed a book,* Nagashizuki: the Japanese Craft of Hand Papermaking, *published by Henry Morris of the Bird and Bull Press, North Hills, Pennsylvania.*

Table of the Finished Paper Yields from the Raw Materials

Stage	Kōzo	Gampi	Mitsumata
Freshly harvested trees (undried)	1000 kg.	1000 kg.	1000 kg.
Freshly stripped kurokawa (undried)	330 kg.		
Kurokawa (dried)	170 kg.	170 kg.	160 kg.
Shirokawa (dried)	90 kg.	80 kg.	75 kg.
Finished Paper	54 kg.	44 kg.	35 kg.

Analysis Table of Japanese Fibers

	Average Fiber Dimensions	Characteristics of Fiber	Characteristics of Paper and its Uses
Gampi	Length (average) 3.2 mm Width 0.02 mm	1. Translucent and has sheen like silk 2. Cohesive power between the thin fibers 3. Strong resistance against harmful insects such as *shimi* (a cloth moth or book worm)	1. Thin and almost transparent, strong even under very moist conditions. 2. Used for copy paper, tracing paper for woodblock printing, or manuscript paper for relief printing. 3. Because of its glossy surface and durability it is very suitable for dyeing, for paper with screened patterns, leafing paper, and paper for gold and silver threaded decoration.
Mitsumata	Length (average) 3.6 mm Width 0.023 mm	1. Relatively short, inferior durability 2. Glossy and dense 3. Has elasticity with a resistance to pulling and bending 4. Strong resistance to *shimi* (harmful insects)	1. The most suitable fiber for machine papermaking. Makes very smooth papers, especially for printing. In dilution, *mitsumata* is added to some currency papers. 2. Better paper is obtained by mixing it with other kinds of fibers. 3. Because of its flat surface, it is suitable for calligraphy.
Kōzo	Length (average) 5.5 mm Width 0.018 mm	1. Very long fiber, very durable 2. Good quality in that the fibers interlace well	1. Durable, has special sheen of kōzo, and the plant grows well in Japan. 2. Most Japanese paper is kōzo paper and is used for *shōji* paper, printing paper, mounting paper, decorated Japanese paper.

MOLD-MADE PAPERMAKING: THE EUROPEAN TRADITION

Michel Joly

Paper had always been made by hand until the end of the 18th century. In fact, it is still made by hand in a few mills in Europe today, such as the Richard de Bas mill in France.

Background and History

Several factors led to the mechanized production of paper at the end of the 18th century. Chief among these were the need for paper in large sheets, the development of paper money, the shortage of skilled labor, and the rise of labor unrest.

These reasons were especially true in France, where wallpaper had become quite fashionable during the reign of Louis XV. The political unrest in France also contributed to machine-produced paper. In addition, skilled workers were led to emigrate to join their Huguenot compatriots who had been exiled to England and Holland following the revocation of the Edict of Nantes in 1685. Finally, the French Revolution had done nothing to improve the behavior of paper workers, who had always been known for their unruly character.

Not surprisingly, it was a Frenchman named Louis Nicholas Robert who invented the papermaking machine in 1792. Two Englishmen named Henry and Sealy Fourdrinier, whose name is associated with the horizontal screen belt type of machine, merely developed a machine based on the patent that Robert obtained in 1798.

The great advantage of Robert's invention lay in its ability to produce an endless strip of paper. Soon thereafter, the invention of the cylinder mold machine would make it possible to increase the production of watermarked paper.

It is inaccurate, however, that an American from Germantown, Pennsylvania, invented the cylinder machine in 1809, as reported by J. N. Stephenson in his book, *Pulp and Paper Manufacture*. That machine, too, was invented by a Frenchman. In fact, the cylinder mold machine is even older than the Fourdrinier machine, according to Monsieur Aribert, former director of the French Papermaking Institute, who has been so kind as to provide me with the following details. As early as 1790, two Frenchmen—Ferdinand Leistenschneider, from a small village in Burgundy named Poncey, and Desestables, from a small village in Normandy named Vire—are believed to have built such a machine. They both patented their invention—Desestables in 1807 and Leistenschneider in 1813.

Leistenschneider presented a paper to the Academy of Lyon in which he pointed out the advantages of the cylinder

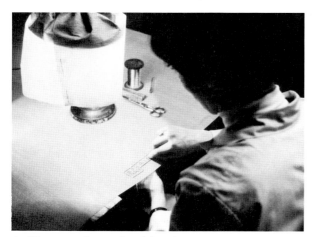

1. Watermarks made of fine brass wire are sewn to the screen. (Photos courtesy of Arjomari-Prioux, Paris, France.)

2. The papermaking screen is soldered, wire by wire, to the cylinder.

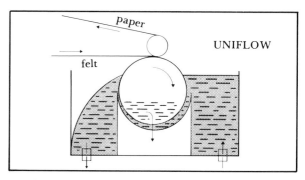

3. The Cylinder Mold Machine. Pulp (dark shade) in dilution moves through the vat. The fibers collect on the rotating cylinder screen as the water (light shade) drains through the screen. The water level inside the cylinder is kept below the level of the pulp solution. The continuous sheet is removed by a moving felt.

mold machine. He said it could equal the production rate of six or seven vats.

Significantly, the cylinder mold machine was inexpensive and required little space. This latter feature is still one of the advantages of the cylinder mold machine over the Fourdrinier type.

In 1805, however, an English mechanic named Bramah filed the first patent for a cylinder machine, which he described as having a very open cylindrical frame, to which a wire screen is attached that gives either a wove or a laid finish to the paper. This frame, which is mounted on a shaft, is open at the ends and has a circular seal that closes its center off from the vat in which it is immersed. The English papermaker Dickinson, who filed a patent in 1809, improved this machine and made it suitable for practical use.

The Mold Machine and How It Works

The modern cylinder machine corresponds rather closely to Bramah's description. It can be covered with a wove screen, that is, with a sort of bronze wire cloth, on which designs can be sewn to produce watermarks (*Fig. 1*). The wove screen can be embossed to produce shaded watermarks. In addition, private watermarks can also be created, custom-made for printshops or for artists who sometimes want their signatures watermarked in the sheet of paper to prevent the counterfeiting of their works.

Likewise, the cylinder mold can be covered with a laid screen, which is made up of relatively widely spaced brass rods joined by more closely spaced laid wires made of the same metal. To produce watermarks, the screen can be die cut, and circular watermarked wires can be sewn in the holes.

The screen, whether of a wove or a laid type, can be mounted on the cylinder by a sewing process. Or, if an invisible joint is desired, it can be soldered on wire by wire (*Fig. 2*). When the cylinder mold is ready, it is immersed in the vat.

The diagram (*Fig. 3*) explains the operating principle of the cylinder mold machine. The pulp (which is dark-shaded) is pumped into the vat in a very diluted form—approximately one pound of dry fiber per thousand pounds of water. This allows the fibers a great deal of mobility. At the same time, water (which is shaded lighter) passes through the cylinder screen as it rotates. Since the water level inside the cylinder is kept lower than the level of the diluted pulp in the vat, fibers collect on the screen. The felt, which a roller presses against the cylinder, takes off the sheet, which is still very wet.

You will note that the process is quite similar to hand papermaking. In particular, as the fibers collect on the screen, they mold themselves around the designs in the screen to produce the watermark. Indeed, its ability to produce fine watermarks is one of the great advantages of the cylinder mold machine.

The fibers can be directed as they collect on the screen by modifying the pulp flow rate, which can be accomplished by lowering the water level inside the cylinder or by varying the speed of the cylinder. Depending on the extent to which the fibers are directed, the paper produced may be more or less isotropic, meaning its physical characteristics will be more or less similar in the grain direction and in the cross direction. For certain special papers, this option is very valuable and is another advantage the cylinder mold machine has over the Fourdrinier.

Naturally, because of the basic principle on which the cylinder mold machine operates, it is not possible to achieve high speeds. In fact, its typical speed ranges from a few meters per minute up to approximately ten meters per minute. In comparison, certain Fourdrinier machines operate at more than a thousand meters per minute.

On account of its very slow speed, paper produced by a cylinder mold machine has a beautifully even formation, an aesthetically pleasing surface, a "friendly feel," and an almost handcrafted appearance. Another advantage of the mold machine is that its relatively small size and slow speed permit the economic production of small orders of special paper, providing a flexibility of size and weight.

Raw Materials

Before continuing this discussion of the next phases of cylinder mold machine papermaking, I believe it would be interesting to tell you something about the raw materials that are used, especially those used in producing papers for art printing.

To obtain long life, among other properties, these papers are made with fibers containing a very high percentage of alpha-cellulose, such as cotton, linen or hemp fibers. Paper-

4. *Spinning mill waste, the raw materials used in making fine art papers.*

makers can produce these fibers from old rags or from spinning mill waste (*Fig. 4*). These raw materials are cut fine and sorted.

In sorting the rags, the papermakers must eliminate any containing optical whiteners because, although these produce an extremely white paper, the whiteness does not last. For this reason, they sort the rags under ultraviolet lighting, under which the optical whiteners appear fluorescent.

Once the rags have been sorted, they are cooked in spherical digesters, fed with pressurized steam and chemicals such as sodium hydroxide and sodium carbonate. After cooking, the rags must be beaten to reduce them to fibers, and bleached. Both of these operations are performed in rag breakers, in which the cooked rags circulate in a continuing stream of water and run between the blades that separate the fibers. To bleach the pulp, the papermakers use a chlorine and hypochlorite solution, which, when drained, produces bright white fibers.

Unfortunately, it is becoming more and more difficult to find old rags free of synthetic fibers, which are not suitable for making paper, at least by these methods. Therefore, it is becoming more common for papermakers to use cotton, flax, and hemp pulps that are produced directly from the plant material, without going through any textile or rag stage. In this case, they obtain raw material in sheet form.

Refining

Whether or not the fibers are of rag origin, they must undergo refining in the paper mill. In this process, the fibers are suspended in water once more, at a concentration of approximately four or five pounds per one hundred pounds of water. The fibers swell and are then refined in beaters.

This type of beater has blades mounted on a cylinder. The blades may be made of bronze, steel, or volcanic stone. The cylinder revolves and comes nearly in contact with a plate which likewise is fitted with blades. Often, these machines are called Hollander beaters, and, indeed, they are similar to the machines developed by the great papermakers in Holland in the 17th century.

The function of these beaters is to crush the fibers, which increases their surfaces for greater contact between fibers and thus stronger felting. Extensive beating makes it possible to develop certain of a paper's physical properties, such as tensile strength, folding strength, and burst strength. Often, however, this is done at the expense of others, such as tear strength, bulk, and opacity. Thus, the papermaker's art is often the art of compromise.

At this stage, the beaten pulp is then sized, meaning that the fibers are made more or less resistant to water absorption. If called for, the pulp may be colored. Sometimes fillers, such as titanium dioxide, are added to improve the paper's opacity, or magnesium carbonate, to give the paper some degree of reserve alkalinity, to improve its longevity.

Production and Finishing

Now the pulp is ready to be turned into paper. The screen is lowered into the vat, which is full of pulp (*Fig. 5*). The pickup felt draws the wet sheet into the machine (*Fig. 6*). The sheet passes between granite and hard rubber rolls to squeeze out water.

5. The papermaker lowers the cylinder into the pulp-filled vat.

6. Close-up of the vat, the rotating cylinder (note the shadow marks on the wove screen), and the rotating felt as it lifts the paper from the screen.

To complete the drying process, the sheet is pressed against steamheated drying cylinders. Finally, the sheet is rewound at the end of the machine.

At this point, the paper is often not completely finished, so for fine printmaking papers, the following finishing process is employed. The paper is usually torn by hand with a wooden slitter blade to produce a handsome deckle edge effect.

Earlier, I said that the paper was sized after the pulp-beating process. For certain papers, such as those intended for water-color painting, this sizing treatment is performed after the paper leaves the machine, sheet by sheet, and then the sheets are air dried.

Another finishing operation involves surface smoothing. This may be a continuous processing, known as calendering, in which the sheet passes between steel- and paper-filled cylinders that glaze the surface. Another method is where the paper may be smoothed sheet by sheet. Each sheet is sandwiched between two cardboard sheets having a very smooth surface. A stack of these is made up, and it is then inserted into a press, where it undergoes compression in a back-and-forth movement. Prior to shipment, the paper is sorted and wrapped.

Even though these papers are virtually handmade, they are routinely tested in our laboratories. Their physical characteristics, such as tensile strength, thickness, and surface finish, are measured. By use of the "xenotest" method, the papermakers can test the paper's resistance to aging and yellowing.

The sensitive papermaker does everything possible so that the paper meets the needs of the artist who seeks to express himself without constraint on a surface that brings out the value of his work and insures its long life.

Michel Joly has worked with Arjomari Prioux paper mills in France since 1961 and currently serves as manager of Research and Development.

HAND PAPERMAKING, U.S.A.

Robert Serpa

While the craft of hand papermaking is ancient and the methods used to produce paper have changed little since its introduction into the Western world, it is a process that is remarkably little understood today, especially when we consider the vast number of times we come in contact with some form of paper in our daily lives. It may be that the tremendous volume of unsolicited paper in its myriad of forms that we have to contend with has dulled our senses and, until recently, dampened our curiosity about the medium.

Happily, however, there has been a resurgence of interest and activity in paper by artists, preservationists, and the public. The current attention being given to the craft is due in no small part to the interest created when the medium has been used as a vehicle to create works of art. Many artists are beginning to see paper as a forceful and exciting medium for their attention and energies.

One institution, among many others, that has been concerned with the archival quality of paper is the Library of Congress. Much to their chagrin, Library employees have discovered that a great many books in their care, produced since the advent of the machine process, are not meeting the test of time. A Library of Congress study of books produced between 1900 and 1939 concluded that 97% had a life span of 50 years or less. According to the estimate of Frazier Poole, the chief preservationist of the Library of Congress, "Present day book paper, as bad as any ever made in history, has a 30 to 35 year life expectancy." While an artist can use the raw materials of papermaking with little or no concern for archival properties, it is very important for the hand papermaker, who intends to provide a substrate for the fine art and book paper markets, to concern himself with these matters.

The opportunity to obtain papers from or work to specification with commercial hand papermakers is readily available today. Though in his seventies, Douglass Howell still produces papers for distribution from his home/shop in Locust Valley, New York. Kathy and John Koller operate HMP Papers from their home/studio in Woodstock Valley, Connecticut. Joe Wilfer divides his time between his Upper U.S. Paper Mill in Oregon, Wisconsin, and his work as director of the Madison Art Center. Kathy and Howard Clark's Twinrocker, Inc., is in Brookston, Indiana, while Farnsworth & Company is presently in San Francisco, and my own Imago Hand Paper Mill is in Oakland, California.

All of these papermakers are concerned with the archival properties of the papers they produce, and each establishment offers the artist the opportunity to work on a custom basis with the papermaker to obtain the best possible sheet for the project at hand.

2. The sheet is formed. The time involved in the actual sheet formation is quite short, only a few seconds. The mold is dipped into the vat and is covered with water and fibers. As the mold is lifted from the solution, the water begins to drain as the vat person sets his shakes, causing the fibers to interlock in all directions so as to produce a square piece of paper. Once the water is sufficiently drained, the vat person places the mold on the bridge of the vat, removes the deckle, and places it on the second mold.

1. Serpa is shown loading a sheet of linter fibers into the Hollander beater.

Paper is made in the beater. Once the papermaker has chosen his materials, the treatment he gives the fibers in the beating process will determine the dominant characteristics of the surface he then creates at the vat.

When the pulp has completed the beating process, it is stored in stuff chests until it is added to the vat. When the vat has been charged with pulp from the stuff chest, the vat must be stirred sufficiently to allow the fibers to become freely suspended in the water.

3. Meredith Mustard is shown finishing her couch (rhymes with "hooch") while Serpa forms a new sheet.

In production papermaking, as the vat person makes the next sheet, the coucher transfers the freshly made sheet to a felted blanket by rolling the sheet off the mold with sufficient pressure to cause the paper to attach to the blanket, yet delicate enough so as to avoid distorting the fibers. The coucher then places a felt on the freshly made sheet. It is the rhythm of the vat person and the coucher that establishes the production of the vat.

4. Once the post of freshly made sheets is completed, it is loaded into a press. The pressing creates a physical bond between the fibers and begins the long and delicate drying process.

5. *Mustard presses the paper. Note the degree of compaction achieved.*

6. *Mustard assists Serpa in registration of the freshly pressed sheets. Once the sheets have been removed from the felts, they will be pressed again with less pressure and left to weep. After the third pressing they will be taken to the drying room or loft and dried.*

7. *Serpa picks the surface of the sheet in grading sheets prior to packaging.*
   Once the completed sheet is dried, it must be inspected and graded in quality of firsts, retrees, and seconds.

Robert Serpa, a professional hand papermaker since 1976, is the owner of the Imago Paper Mill in Oakland, California.

CHEMISTRY OF PAPER

Roy P. Whitney

I'm going to tell you something about the materials used in making paper and to explain why some things happen and what can be done to alter the results. If I can accomplish this, perhaps you will have a little better understanding of the materials papermakers work with. I'll try to be as nontechnical as possible, although that may prove to be a bit difficult at times.

Natural Fibers

Paper is a sheet material made by bonding together many very small discrete elements called fibers. It seems obvious that we share a common interest in fibers, so let's talk about them for a few minutes.

We say that a material is "fibrous" if its elements are slender, threadlike, and filamentous, that is, one dimension is very much greater than the other two. Fibers may be composed of animal, vegetable, mineral, or synthetic materials. All are used in papermaking, but we are concerned particularly with vegetable fibers. They are usually called natural fibers, because they grow in nature.

Natural fibers can be classified according to the manner in which they grow, as well as their location and function in the plant or tree. One class that is particularly important is the "seed-hair" fiber. Cotton is a seed-hair fiber.

I'm sure most of you have seen cotton in bloom and have seen the beautiful bolls, made up of very long staple cotton fibers. Many fibers are attached to each seed, and the fibers can be as much as 1½ inches long. These very long fibers go almost wholly into textiles and are not used in paper except as old rags, which used to be an important papermaking material.

We still hear the term "rag content" used, and rags are still used sometimes, particularly in handmade papers. But when the staple cotton is separated from the seeds in the cotton gin, shorter fibers remain and are recovered separately. These are called "cotton linters," and they are an excellent papermaking material for specialty papers such as high grade bonds and art papers, among others. Linters fibers can be up to five or six millimeters long, or nearly one-fourth inch.

Among the stem or trunk fibers are several that are important in papermaking. One is the "bast" fiber. It is found in the inner bark of numerous shrubs and trees, and it was the first fiber used in papermaking. Flax is a familiar bast fiber and was probably the first plant fiber used in textiles (linen).

Others, used particularly in the Orient, are mulberry and mitsumata.

The trunk fibers of both softwood and hardwood trees provide the principal source of fibers for the commercial paper industry today. Depending upon the wood species, they vary greatly in length, from about one-half millimeter to five or six millimeters. Wood provides a renewable source of raw material, without which the paper industry as we know it would not exist.

Straw is also a source of fibers from which excellent papers are made, but its use in this country has diminished steadily for many years.

Unfortunately, we cannot discuss fibers to any extent without getting into cellulose chemistry, which is apt to be a bit technical. I won't say much about it, but I do think you must at least be aware of cellulose and its behavior.

Cellulose is the basis of all natural fibers. It is the fundamental building block from which fibers are formed. Cellulose is a product of photosynthesis, one of the most ancient of processes that go forward on our planet. It is the most abundant organic material on earth, with an annual growth rate estimated at one hundred billion tons per year, far exceeding the growth or production of any other material.

Chemically speaking, cellulose is a polymer of glucose, which is a common sugar. Now we all know that "poly" means "many," and a chemical polymer is a large molecule made up by bonding together many smaller molecules. So, in cellulose, we have a molecule consisting of many repeating units of glucose. The process of building up a polymer can proceed in different ways. For example, the bonding may take place in a random manner in all directions, to form a three dimensional structure.

It doesn't do that in cellulose, however, which is a straight-chain polymer. That is, each new glucose unit adds on to form a chain, which can be represented much like a long string of beads. The cellulose chain may contain anywhere from a few hundred glucose units to as many as several thousand. It is dangerous to generalize, but often the longer the chain, the more resistant is the cellulose to degradation and deterioration. Cotton, for example, is a very stable and resistant fiber, and it has a very long chain length.

Cellulose and glucose are carbohydrates, which means that they are composed of carbon, hydrogen, and oxygen, with the hydrogen and oxygen atoms always present in the same ratio as in water (H_2O). This is no accident, but results from the evolutionary process on our planet. If water were not so abundant on earth, we would not have cellulose. Cellulose occurs principally in nature as the hollow elongated biological cell found in all plant life, which we call the fiber.

Cellulose has a great affinity for water, so we say it is "hydrophilic." Cellulose molecules also have a great affinity for each other, particularly when they are wet. If two cellulose chains are brought into close proximity, they bond together very tenaciously. And this is how the fibers are built up.

Cellulose chains bond together to form sub-units called "fibrils," and these bond together to form fibers. This is all done in a very systematic way as the plant grows, and the molecules and fibrils and fibers all create a remarkably ordered system. If one dissects a cellulose fiber, he finds various layers of fibrils in the cell wall, each wound helically around the fiber. For a given plant, each set of windings is at the same angle with respect to the fiber axis.

All plant fibers are hollow. They have an empty core inside the cell wall, which is called the "lumen." In various fibers, the lumens may be large or small, and conversely, the cell walls may be thin or thick. As we shall see, the relative size of the lumen and the thickness of the cell wall is very important in papermaking.

Of course, cellulose is not the only chemical component in plants and trees. Among others, there are minerals, which provide nutrients during growth, and there are resins and gums, generally called extractives. We usually get rid of most of these before forming the fibers into paper. If we disregard them, we are left with three major constituents—cellulose, hemicelluloses, and lignin. Most chemists believe that all plants and trees contain some of each of these substances, although the proportions vary widely.

Hemicelluloses are very similar to cellulose in composition. They are straight-chain polymers of sugars other than glucose, but the chain length is usually shorter than cellulose—a few hundred repeating units rather than a few thousand. They are generally much less resistant to degradation, and to attack by chemicals or by atmospheric conditions. Hemicelluloses bond even more readily than cellulose; in fact so readily that they sometimes form sheets, which are much too dense and translucent. So, depending on what kind of paper we are making, we sometimes try to retain the hemicelluloses in the fibers and sometimes try to eliminate them.

Lignin is often called the cement that glues the fibers together in the shrub or tree and gives it the structural strength to stand straight and grow tall. Lignin is a three-dimensional polymer, it is amorphous, and it has no ordered structure. It is not fibrous. In contrast with cellulose, it doesn't love water, it hates it, just as does a petroleum product. So we call it "hydrophobic."

Unlike cellulose, lignin does not have a specific chemical composition or structure. It varies considerably among plants. So we cannot speak generically of lignin, but we should use the plural, "lignins," to indicate this. Lignins are undesirable materials in papermaking, and if we are making fine papers we try to get rid of them to the greatest extent possible.

So far, I've noted that the proportion of these three major components varies greatly among various trees and shrubs, and the following table indicates the extent of this variation. I've picked four plants which should be particularly interesting to you. It must be emphasized that these percentages are only approximate and average. Specific analyses can be found that differ appreciably.

These figures should indicate to you one reason why cotton is such a good papermaking material. It is the purest form of

Approximate Composition
(Extractive-free Basis)

	Cotton Linters	Raw Flax	Typical Softwood	Typical Hardwood
Cellulose	96%	85%	50%	50%
Hemicelluloses	3%	10%	20%	30%
Lignins	1%	5%	30%	20%

cellulose occurring in nature, and hence not much pulping or purification is necessary before the fibers are used. Flax and the other bast fibers are not quite as good, but nevertheless they are attractive. We can derive excellent papermaking fibers from wood, but extensive pulping and purification are necessary, and this is expensive. The great virtue of wood is its continuing availability in tremendous quantities.

Fiber Morphology

Although it is not appropriate to become deeply involved in fiber morphology, perhaps a few pictures of wood sections will illustrate some of the points I have been trying to make. *Figure 1* is a scanning electron micrograph of a section of jack pine, showing both the radial and the transverse planes. The radial plane, near the top of the picture, is cut horizontally across the tree trunk, and shows the cross sections of the vertical fibers. The transverse plane, near the bottom, is cut lengthwise of the trunk and of the fibers.

I think you will agree that this is an example of order of a very high degree. The cells are roughly square in cross section. The cell walls are very thin, and the lumens are therefore large. In other words, these fibers are mostly hollow cores.

These are "springwood" or "early wood" fibers, formed in the spring of the year when the growth rate was high, and such fibers are always characterized by thin walls. In the transverse section, the pits in the fiber walls are evident. Pits are characteristic of soft wood fibers, and they form connecting horizontal passages between fibers to permit fluids to be transported up the tree through the lumens.

If these fibers are isolated and subjected to beating and violent treatment, as they are in Oriental hand beating or in a Hollander beater, it might be expected that they would collapse and look much more like ribbons than filled-out fibers. This is exactly what happens in papermaking with typical springwood fibers, particularly from softwoods.

Springwood fibers are very thin-walled as compared with thick-walled "summerwood" or "latewood" fibers. As the growing season progresses, the rate of growth slows appreciably, the fibers develop very thick walls, and correspondingly the lumens become smaller. In contrast with spring-

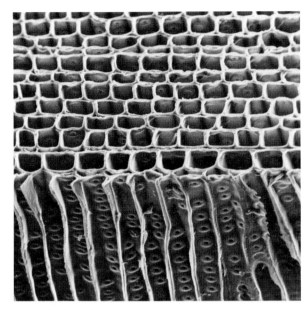

1. *Radial and transverse planes of jack pine. Scanning Electron Micrograph (SEM), magnified 160 times. (This and following micrographs are taken with permission from* Papermaking Materials — An Atlas of Electron Micrographs *by R.A. Parham and Hilkka Kaustinen, The Institute of Paper Chemistry, Appleton, Wisconsin, 1974.)*

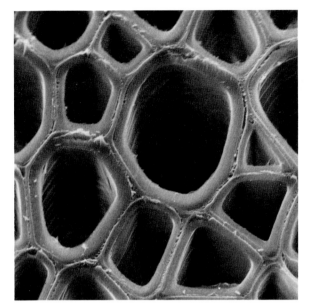

2. *Fiber cross sections of aspen, a hardwood. SEM, magnified 1760 times.*

wood fibers, the summerwood fibers are much sturdier, and for the most part they do not collapse on beating.

Similar sections of hardwoods show growth rings, and the same transition from springwood to summerwood. Hardwoods are more complex structurally than softwoods—they have progressed further along the evolutionary chain. They do not have pits but employ a different mechanism for transporting fluids up through the trunk. In addition to the normal trunk fibers, hardwoods have vessels, relatively large conduits that extend from the bottom to the top of the tree. As we shall see, hardwood fibers are much smaller than softwoods, and they do not collapse as readily.

We spoke a moment ago about the three major com-

ponents of all trees and shrubs, but we said nothing about their location in the structure. *Figure 2* is a radial section showing fiber cross sections in aspen, which is a hardwood, at a much higher magnification. As you see, these are typical springwood fibers, with very thin walls and large lumens. Essentially, all of the cellulose and the hemicelluloses are in the cell walls. But I'd particularly like you to look at the boundaries between the walls of adjacent fibers and the rather ragged and ill-defined material located there. This is lignin, and this area is the "middle lamella." This lignin indeed serves as a cement to stick the whole structure together, but unfortunately, not all of the lignin is located here. About half of the lignin is in the middle lamella, and the other half is interspersed throughout the cell wall. I say "unfortunately" because the middle lamella lignin is relatively easy to remove in pulping. The cell-wall lignin is very difficult to remove and cannot be taken out without some degradation of the cellulose.

It has been noted that fibers vary greatly in size. Indeed, they range from five millimeter cotton linters downward through the hardwood fibers, with the maple fiber being one-half millimeter in length. It seems evident that these great differences in size should cause differences in the properties of papers, and that is indeed the case. Many grades of paper contain blends of fibers in order to take advantage of the contributions of each.

Most natural fibers are tapered at both ends. An exception is cotton, which has its greatest diameter at the end that was attached to the seed, and it tapers over its whole length to the tip.

Pulping and Bleaching

I think it is not necessary to spend a great deal of time on the papermaking process, as it is well-illustrated elsewhere in this book. I'll try to confine myself to showing you what happens in the various processing steps.

We start with a source of fibers, which might be raw cotton linters, the inner bark of the mulberry tree, or part of a tree. With all materials except cotton, we must liberate and separate the fibers, and this is one of the major objects of pulping. All fibers are pulped, either mechanically or chemically or both. I sometimes get the feeling that this is not generally realized, and that there is an impression that pulping is detrimental and to be avoided. It may be detrimental if not properly done, but it cannot be avoided.

If the pulping is solely mechanical, as in making groundwood, all of the lignin remains in the pulp, and the resulting papers are of very poor quality. They have poor strength, they yellow rapidly on aging, and they are used in products where quality and long life are not important, such as in newsprint and the groundwood printing paper used in most paperback books.

In chemical pulping, the object is not only to liberate the fibers but to purify them. Numerous chemical reagents are used, and the pulping can be mild or quite drastic. The cotton

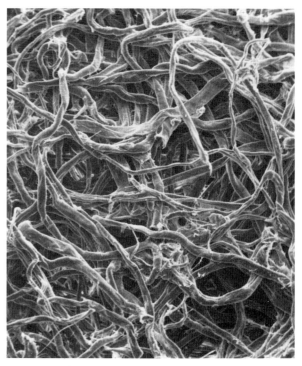

3. Cotton linters fibers. SEM, magnified 115 times.

linters that you buy have had a mild alkaline pulping to remove gums, waxes, and some other noncellulosic materials. Pulp used in Japanese hand papermaking is steamed to release the inner bark from the other components, and then pulped by boiling for two hours with sodium carbonate, an alkali, to liberate the fibers. Wood, because of the high lignin content that must be removed, is subjected to more drastic chemical pulping treatments.

So, all of our fibers have undergone a pulping treatment, of greater or lesser severity, and with more or less degradation of the cellulose fibers. Properly conducted, the pulping need not seriously degrade the fibers.

Bleaching can be considered a continuation of pulping, to purify further and to whiten the fibers. Particularly with wood

4. *Bleached* mitsumata *fibers. SEM, magnified 115 times.*

After pulping and bleaching, the next major process step is beating and refining. Beating is always necessary to make good paper. The dimensions and the surface of the cellulose fiber must be controlled to obtain the desired properties.

Two major effects of beating are cutting and fibrillation. Cutting results in shortening of the average fiber length and is necessary particularly with long fibers to obtain good sheet formation. Very long and very flexible fibers give sheets with "wild" formation. That is, the fibers agglomerate or flocculate into clumps that show up as thick places in the sheet, with thin areas surrounding them. I'm sure you have all observed the mottled appearance of some papers when viewed against a strong light. This tendency toward flocculation is controlled largely by adjusting the fiber length. Of course, cutting can be carried to the extreme of impairing the fibrous nature of the pulp and ruining its papermaking qualities.

Fibrillation involves altering the surface of the fibers by macerating, fraying, brooming, and generally causing the intricate fibrillar structure to be disrupted. This creates much more surface, which is then available for bonding in the papermaking step. Beaten fibers are much less well defined than unbeaten ones, but they bond together much better and produce stronger, smoother, and generally better paper.

Figures 5 and *6* show the same softwood fibers before and after beating. The beaten fibers are more collapsed, more transparent, and they form a more compact mat with less void space.

Most of what I have said about beating is summarized in *Figure 7.* The development of surface, the freeing of the surface fibrils, and the interfiber bonding which results are strikingly illustrated. The old papermakers were certainly right when they used to say that "paper is made in the beater."

The next major process steps are sheet forming and drying, and these have been covered in detail elsewhere in this book. You know that essentially all paper is made by a filtration process. A very dilute slurry of fibers in water is prepared and caused to flow onto or over a porous screen. The water drains through the screen, depositing the fibers as a mat. This wet mat is removed from the screen, water is pressed out, and the

5. *An unbeaten kraft softwood pulp. SEM, magnified 47 times.*

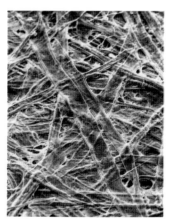

6. *Same pulp as* Figure 5, *after beating. SEM, magnified 47 times.*

pulp, bleaching is effective in removing much of the lignin in the cell wall. Most bleaching today is conducted with chlorine compounds, and it must be done very carefully to avoid cellulose degradation. Like pulping, bleaching is a necessary operation, but it must be conducted with care and understanding of what is going on.

Cotton linters are shown in *Figure 3,* another scanning electron micrograph. Cotton fibers are thin-walled, with large lumens, and they collapse readily into ribbons. They also have a tendency to twist about their long axis, as some of these fibers have done.

Figure 4 shows *mitsumata,* a bast fiber, at the same magnification. As you see, these fibers are much more slender than the linters, and this makes them more flexible.

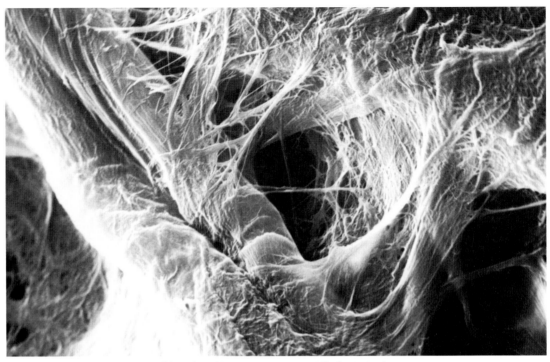

7. Interfiber bonding. SEM, magnified 2250 times.

sheet is dried. The equipment used has been developed considerably over the past 18 centuries, but the basic process remains the same.

Since about 1800, two types of continuous paper machines have been in commercial use. These are the cylinder machine and the Fourdrinier. The cylinder machine consists essentially of a wire-covered cylinder rotated in a vat containing the fiber slurry. The sheet is formed on the cylinder surface, couched off, pressed, and dried. The Fourdrinier comprises an endless wire screen, with the pulp slurry flowing onto it at one end and the wet sheet couched off at the other. The pressing and drying operations are essentially the same as in the cylinder machine.

The cylinder machine has serious limitations as regards control of sheet properties and speed, and is essentially obsolete for commercial purposes. It is still used in special situations, one such being the production of art papers. The Fourdrinier is much faster and more versatile, and it dominates the commercial scene. However, it shows signs of obsolescence, too, and other forming devices are coming into use. They still, however, employ the original concept of drainage of water through a porous screen.

Chemical Additives

It would be difficult to overstate the importance of chemical additives in papermaking. It is safe to say that a large fraction of the 60 million tons of paper produced annually in the United States could not meet the varied end use requirements without the additives which impart the desired sheet properties. Paper can be made water-absorbent or water-resistant; it can be filled, dyed, coated, impregnated; it can have remarkable wet strength or almost none at all. These and many other characteristics are made possible through the use of chemical additives.

Internal sizing has long been employed to control wetting of the sheet and to permit the paper to accept ink and other aqueous fluids without undue feathering or spreading. The principal internal sizing material is rosin. The rosin is cooked with an alkali, or otherwise treated, to form a paste that is added to the furnish before the paper machine. Alum is then added, to precipitate very fine rosin particles on the fibers and render them water-resistant. However, a difficulty arises in that alum is a very acidic material, and the whole papermaking system as well as the finished paper becomes acid as a result.

This may be a good point to digress and to speak briefly about acidity and alkalinity and pH. Water ionizes into what are called hydrogen ions and hydroxyl ions. By some mathematical manipulation and by knowing the ionization constant, we can show that the sum of the logarithms of the reciprocals of these two ionic concentrations is constant at 14. The hydrogen ion concentration is a measure of acidity, and the logarithm of the reciprocal of the hydrogen ion concentration is the pH.

If this all sounds pretty complicated, just remember that pH is a measure of acidity on a scale of 0 to 14. At a pH of 7, the solution is neutral, with equal numbers of hydrogen and hydroxyl ions. As the pH moves progressively below 7, the solution becomes more and more acidic. Conversely, as the pH moves above 7, the solution becomes more alkaline. But remember also that this is a logarithmic scale, and therefore very sensitive. A solution of pH 5 has 10 times the acidity (hydrogen ions) of one at pH 6. At pH 4, the solution is 100 times more acid than at pH 6.

Most rosin-sized papers are made in the pH range of 4½ to 5½. We know now that this is much too acid for the papers to show any great degree of permanence, and they will deteriorate significantly in a few years. Papers of archival quality must be made at essentially neutral pH. In fact, they often contain alkaline materials that neutralize any acids that may form over years of storage.

Fortunately, other internal sizing agents are now available that can be used under neutral or slightly alkaline conditions. One particular agent is called Aquapel. It is a synthetic

material and a good sizing agent. So, if you want your papers to last a long time, you should be careful about rosin sizing and particularly about the amount of alum you use.

Surface sizing is applied after the sheet is formed and dried, and it is therefore mostly on or near the surface. It is used to impart surface finish and to control surface properties. Quite different materials are used—materials such as glue, gelatin, casein, and starch. Most bonds and writing papers, and many art papers, are surface sized.

Filling and loading are also practiced extensively with many papers. Fillers improve the opacity and also provide further control of surface properties. Most fillers are minerals that have been very finely ground. Kaolin clay is the most common filler. Calcium carbonate (ground limestone, precipitated chalk) is also used extensively in alkaline papers. Titanium dioxide is a preferred filler, but it is also about the most expensive. Some synthetic materials are used.

Fillers always weaken the sheet. Retention of fillers is also poor, since they have no affinity for cellulose and wash out readily during the sheet forming process. A well-closed white water system is essential for good filler retention.

Coloring or dyeing is an essential part of papermaking. Practically all papers are dyed, even all white papers. In fact, white papers are just about the most difficult to match.

Most paper dyes are synthetic, organic, water-soluble materials. Some pigment dyes are used, and they behave like other fillers. There are three classes of dyes in use commercially. These are direct dyes, which have a strong affinity for cellulose and are probably used the most; basic dyes, with a strong affinity for lignified (and therefore unbleached) fibers and are used widely in groundwood papers; and acid dyes, with no affinity for cellulose. Acid dyes require a mordant to set them, and the usual mordant is rosin and alum, which we have already discussed.

Among other chemical additives are beater adhesives, such as starch, natural gums, modified celluloses, and wet-strength resins. Some of the natural gums and mucilages have remarkable abilities to control fiber flocculation and sheet formation. I've only scratched the surface of what might be said about additives, and if you are a papermaker, this is a fertile field to explore.

Paper Properties

Let's wind up this rather long discourse with a few comments about paper properties. We have not yet discussed the strength properties of paper, for example. I doubt that paper strength is of major concern to you, but nevertheless, you should be aware of the effect of processing variables on strength.

In *Figure 8*, I have plotted three common strength properties against the degree of beating. As you see, the sheet tensile strength increases with beating time until a maximum is reached, after which further beating is detrimental. The bursting strength, which depends upon both tensile and stretch, shows about the same pattern as does tensile. The tearing resistance shows a quite different trend, however, and generally decreases as tensile increases. Usually, increasing tensile and burst indicates greater bonding of the fibers, and this usually decreases the ability of the sheet to withstand tearing.

Sheet density is also heavily dependent upon the degree of beating. The greater the extent of fiber bonding, the more compact and dense will be the sheet. Extremes in density might be a sheet of blotting paper, which is very lightly bonded, and a glassine sheet, which is so thoroughly bonded as to become translucent.

Some sheet properties depend upon the paper machine and the way it is operated. An example is two-sidedness. All sheets have a wire side and a felt side. The difference is quite pronounced in some instances. Also, machine-made papers show uneven fiber orientation, with quite different properties in the machine direction and the cross direction. Handmade papers should show little, if any, directionality.

Certain sheet properties also depend significantly on the moisture content, which in turn is governed by the relative humidity of the air in which the paper is stored. The moisture content of paper may be almost zero in very dry air and as much as 30% at close to 100% relative humidity. Paper also changes dimensions with moisture content, a property called hygroexpansivity. Fibers expand and contract principally in cross section rather than length. So if the sheet has a high degree of directionality, the dimensional change will be greater in the cross direction than in the machine direction.

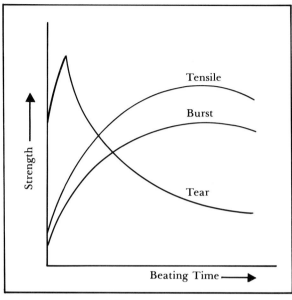

8. Development of Sheet Strength with Beating.

Let me close by saying that I think it is wonderful that so many people have rediscovered an old art, and are enjoying it and finding it useful. However, I have a few words of advice to those involved in hand papermaking:

Please don't fail to take advantage of all that has been learned about papermaking in the last century or so. Don't focus your attention exclusively on cotton or flax or any other single fiber, but learn about the flexibility and versatility that can be achieved with blends of fibers. And please remember that you must do more than start with fine fibers. You must process them in such a way as to enhance their potential and avoid damaging them.

Dr. Roy P. Whitney, a chemical engineer, has served on the faculties of the Massachusetts Institute of Technology, the University of Maine, and The Institute of Paper Chemistry. He has published numerous papers and is currently assistant to the president of the Institute of Paper Chemistry, Appleton, Wisconsin.

LABORATORY PAPER TESTING

Donald Farnsworth

Most of us at one time or another have tried to bring some intelligence to the selection of a paper for our use. Occasionally, we rely on the paper manufacturer to provide us with information about the use for which the paper was designed, and it is frequently described in its name: German Etching, Simpson Offset, Archival Parchment, or Venetia Watercolor paper. However, these descriptions are merely generic and tell us little about the functionality of the paper for our particular needs.

Lyrical titles, such as Stonehenge, Perusia, Umbria, or Rosapina are even less helpful although they do capture some of the romance long associated with paper. Manufacturers usually provide literature recommending uses for the paper to promote its sale, but often the paper has not performed in the studio in the manner we hoped. Just as often, we have found uses for it that were unexpected.

While the problem of a paper's use is complex and sometimes despairs of simple answers, we are nonetheless equipped with some resources of our own to help us make an informed decision. Our preeminent resource, of course, is our own judgment and our five natural senses. We visually examine each paper that interests us; we feel its texture, weight, and stiffness; we listen to its rattle; we observe its color and the way it reflects light, noting its gloss, its opacity, its torn and cut edges. By this simple examination, we will have performed some elementary testing of paper.

The investigation will not progress long before we discover that the papers available to our scrutiny all differ in character in great and subtle ways—from very soft and flexible tissues to rigid museum board, from slick, highly calendered to pebbly, highly textured papers. By touching them and looking at them reflect light, we can discover something about the surface texture of the papers. The contour or exterior of paper is said to have a "finish" or "pattern," which is "smooth" or "rough."

By bending the papers between our fingers, we can ascertain informaion about its rigidity. We usually consider rigidity, or the paper's resistance to being bent, desirable in a paper. By holding one edge of a sheet and extending it horizontally, we will be able to tell whether or not a paper will support its own weight, and thereby learn something of the stiffness of the paper. Stiffness, rather than limpness, is often considered a sign of a paper's quality. By pressing the paper between our fingers, we might inspect the paper's sponginess (or compressibility), as well as its soft feeling.

Although we can obtain a considerable amount of information this way, other important features of each of the papers will not be revealed to our unaided senses: durability for one (or resistance to the destructive effects of abuse, handling, environment, and time); acid content or pH for another; dimensional stability (that is, the tendency to change dimension in response to changes in moisture) for a third. While it may seem empirically obvious that the thicker, heavier sheets will be stiff and rigid while light and thin paper will be soft and flexible, this is not always the case. Blotter paper, for example, can be thick and heavy yet pliable and soft. Glassine or bank note paper can be at once light and thin as well as hard. While it may be inviting to assume that dense papers are necessarily stiff, this too is not always so. Indeed, within a single sheet of paper several contradictory properties may exist.

Finally, unless the art store allows its customers to pull apart its samples or tear off snippets from its stock, we will have no indication of the paper's relative tensile or tear strength. To obtain information such as this, we would have to take the papers to a laboratory and scrutinize them under more controlled conditions than are available in an art store.

Most of the information we might obtain from extemporaneous experiments in the art store would be thoroughly subjective and, therefore, only as good as our perceptions at that time. The varieties and the properties of paper that exist are so numerous that the identification and expression of these differences require the application of standard test methods.

Tappi Standards

Consequently, a definitive body of standards for the testing of paper, known as the Tappi Standards, has been devised for the papermaking industry by an organization known as the Technical Association of the Pulp and Paper Industry. Most laboratories and large paper mills in the United States recognize Tappi Standards, and most paper testing equipment is calibrated and used according to those references. So to extend our knowledge of paper beyond the superficial properties we can discover in the store, we have available a standardized language by which we can communicate our knowledge with each other. However, our understanding of its vocabulary requires

of us at the very least a fundamental understanding of the chemistry of paper.

As almost anyone concerned with the subject of paper should know, the basic material of paper is cellulose. Cellulose is a polymer of glucose units strung together in long chains that tend to lie parallel to one another in a regular array. Found in most plant life, cellulose fibers used for papermaking are most commonly derived from wood, rags, cotton linters, straw, mulberry, and flax. Many millions of these fibers are matted together to form a single sheet of paper, but they first must be pulped—the bonds of the cellulose fiber must be ruptured—and dispersed in water.

As cellulose fibers are physically bruised by beating in water, they swell, a reaction known as hydration. Cellulose fibers have a high self-bonding, sheet-forming ability if they are brought into intimate contact with one another. So, the increase in the surface area of the frayed and swollen fibers, caused by beating, increases the fiber-to-fiber bonding. Thus, hydration makes dense paper and hydration carried on for extended periods reduces opacity and will eventually make a stiff, rattly, translucent paper. A heavily hydrated pulp may form nearly transparent paper. A high degree of hydration makes paper dense and hard, and it provides a cohesive strength that will produce in the laboratory high results from a laboratory Mullen test.

Testing Paper Strength

The Mullen paper tester is one of several recognized devices for testing the "strength" of paper. In Mullen's apparatus, the paper specimen is clamped over a circular diaphragm of thin rubber. By turning a wheel, a piston forces glycerine against the underside of the diaphragm, which presses against the paper with increasing force until the fibers are pulled apart (*Fig. 1*). The pressure at bursting point is recorded on a dial by a pointer. A Mullen test generally indicates how long a pulp has been beaten and how high or low its quality. The higher the number recorded on the dial, the more the resistance to bursting is indicated.

A bulky sheet of paper can be produced by using pulp

1. The Mullen Tester records the bursting strength of paper.

barely agitated in water — a minimum of refinement — for a very brief time, then forming as thick a sheet as possible and pressed with a minimum of pressure. Such a paper would be coarse, fuzzy, weak, and rough. Even lacking good sheet formation, its thickness alone would endow it with substantial resistance to bursting. However, a thin sheet of high quality, well-hydrated pulp could easily be fashioned to surpass the bulky paper in a Mullen test. The bulky paper would show much higher in the Mullen test than would a thin sheet from the same poorly refined pulp. That is why one should calculate measurements from strength tests against the basis weight of the paper.

The basis weight simply indicates the weight of the paper as it would appear in a "standard" ream of 500 sheets measuring 38 inches by 25 inches. The more dense and thick each sheet of paper in the ream, the heavier the ream.

Resistance to Tear

Thickness and denseness can be said to improve bursting strength but not necessarily resistance to tear. The longer the fibers used in forming the paper, the higher will be its resistance to tear. However, short fibers seem to form better than long fibers and also hydrate more rapidly. More hydration means the creation of a dense, hard, rattly sheet of paper, but papers formed from short fibers usually show low results in tests for tear resistance.

A generally acknowledged test for measuring the resistance of paper to tearing action is the Elmendorf test. In this experiment, a piece of paper (or several thin pieces) is fixed between two clamps close together, one being fixed and the other attached to the lower end of a pendulum device. The specimen is slit so that only a known distance (63 mm.) must still be torn to separate the piece into two. When the pendulum is released the paper is torn between the fixed and the moveable clamp. The greater the resistance (to tearing) of the paper, the more the swing of the pendulum is impeded and brought to a standstill, which is indicated by a friction pointer. The pointer indicates a number on a scale, which is the measure of the "tearing resistance." The higher the number indicates the greater the paper's resistance to tearing.

In machine-made papers, two such tests are usually performed. One tests the resistance to tear in the "machine direction" (MD) or the "grain" of the paper, that is, the direction the paper was coming off the roll. Since the cellulose fibers generally orient themselves parallel to the machine direction of the paper, the machine direction should show lower resistance to tear, as tearing would more naturally occur along the axis of the fiber arrangement. The other tests resistance to tear in the "cross direction" (CD) of the paper, or the direction across the sheet (against the "grain"). One would expect the cross direction to show higher resistance to tear than the machine direction of the paper since tearing would be attempted across the natural alignment of the fibers.

Low resistance to tear would be expected in papers formed from short fibers, from underbeaten pulp, or fibers that have been damaged by overbeating. The condition can be improved by the addition of long fibers to the pulp. Long-fibered pulp that is moderately hydrated can form very strong paper. The addition of bonding agents — certain starches, glues, or resins — will slightly improve tear strength and greatly improve tensile strength by increasing inter-fiber bonding.

2. The tensile strength of paper is measured.

Tensile Strength

Measuring the tensile strength of a paper is still another way of measuring the relative bonding between the fibers. This is done by an instrument devised to ascertain what strain, expressed in pounds, is required to break a strip of paper a given length and width. Standard-sized strips cut from the paper in both machine direction and cross direction are attached (in separate tests for each direction) top and bottom to clamps or grips in the machine. The lower clamp is pulled down at a slow, even speed by an electric motor. The strip of paper pulls down the clamp to which the upper end is attached, transmitting the movement along an arm to a weighted lever, which is pulled along a scale.

When the test strip breaks, the lever remains fixed and indicates on the scale the force in pounds required to break the paper (*Fig. 2*). The tensile strength of a paper is described by one number indicating the pressure required to pull the fibers apart in the machine direction and another number for the cross direction. Since tensile strength is tested by a pulling action (instead of a tearing action or a bursting action), the results should be higher in the machine direction than in the cross direction.

Papers that measure a high tensile strength will commonly have been formed from well-beaten pulp because beating maximizes inter-fiber bonding. We would expect them to be relatively dense, hard, somewhat translucent, rattly papers with relatively low tear strength. A paper can be manufactured with both high tear and high tensile strength by mixing long fibers with a well-beaten, hydrated pulp, relying on the long fibers for resistance to tear and the well-beaten pulp to create the bonding necessary for tensile strength. Long fibers, which supply strength in terms of tear resistance and folding endurance, do not form sheets as well as short fibers. Sheets formed with a preponderance of long fiber tend to have a cloudy or "bunchy" look to them.

Folding Endurance

Folding provides a means of measuring the built-in ruggedness and permanence of paper. A paper's resistance to handling and abuse, its durability, can be evaluated from a standardized test known as the M. I. T. folding endurance test. For this test, a half-inch strip of paper under one-half kilogram of pressure is folded a double fold in a 270° arc along the same crease over and over again until the strip breaks (*Fig. 3*). The folding endurance is the number of folds a paper will withstand before breaking under these conditions. A paper will tend to fold more readily without rupturing the fiber bonds along the machine direction than across the grain.

Folding endurance is considered a sensitive measure of the permanence of paper, or the ability of paper to withstand heat, light, internal and external chemical attack, aging, and other influences that tend to deteriorate the fiber. One of the ways such information is obtained is by subjecting test samples to heat and dryness for a specified number of days, and then measuring any loss of strength by comparing the paper's folding endurance before and after this artificial heat aging.

Extremely durable papers are normally manufactured with high-quality rag pulps or long-fibered sulphate or alpha pulp (highly purified pulp derived from wood), or combinations of these. They must be refined well to maintain strength and processed for a long period of time in such a way as to increase hydration without severely shortening fiber length. The result is good tear strength and folding endurance.

Possibly the best known examples of such papers are currency notes. In the United States new rag cuttings are used to manufacture paper for banknotes. New rags provide paper pulp with a high percentage of staple fibers (the longer fibers of the flax stock and the cotton ball) for good folding endurance. A "hard" (chemical) sizing added to the pulp contributes water resistance to the currency, while a starch or gelatin size supplements the fiber bonding.

All of the properties described by these testing procedures concern the ruggedness or durability of paper: bursting strength, tear strength, tensile strength, folding endurance, and aging. Myriad other specialized tests are performed in the laboratory on paper to discover its response to other pressures. This is not to say that the characteristics of paper so tested exist solely in terms of the tests that are performed in order to define them. These properties have direct consequences for us as we consider our choices of paper to use.

Paper Acidity

Thousands of examples could be documented of works of art on paper lost because of excessive internal acid in the paper, or from attack from airborne sulphur dioxide. Since the middle of the 19th century, many printers have inadvertently used commercial papers sized with rosin that they fixed to the fibers with papermakers' alum. Acidic papers, like papers made from poorly purified wood pulps, and papers containing alum-rosin size can be neutralized by today's conservators. However, the damage already caused by these acids is irreversible.

Much discussion has been spent on the issue of paper acidity. Probably acidity is not the problem it once was. At the very least, sufficient varieties of high quality art and book papers are manufactured today so that acidity should not trouble anyone seriously concerned about the permanence of his or her work. In the matter of paper testing, acidity is most often expressed by specifying the concentration of positive hydrogen ions, or pH, in the paper on a scale of one to 14. A pH of 7.0 is considered neutral; paper with a pH of less than 7.0 would be acidic, while paper with a pH of more than 7.0

would be alkaline. Paper with a pH in either extreme can be expected to have a relatively short life expectancy because the excess acidity or alkalinity will degrade the cellulose fibers. This process can be further aggravated by humidity, pollution, heat, light, and abuse.

Testing the surface pH of a paper with commercially available pH pencils is not an accurate method for determining the pH value of the paper. The more accurate and widely accepted method requires boiling one gram of paper in distilled water for one hour before testing the solution with a pH meter. This method is referred to as the "hot extraction method." Buffering agents are now commonly added to paper to neutralize the pH and protect the paper against external acid attack.

Donald Farnsworth is the owner of Farnsworth and Company Papermakers, in San Francisco. He is an artist as well as instructor of papermaking at the California College of Arts and Crafts in Oakland, California. Over the years he has collaborated with many artists at his mill and has developed ways to work with paper as a medium for art.

3. The folding endurance of paper is measured here.

Testing Comparison of Newsprint vs. 100% Rag

	Basis weight (lbs. per 500 sheets)	Caliper (thickness in 1/1000 of an inch)	pH	Mullen (burst in lbs.)	Tear (MD/CD)	Smoothness (felt/wire)	Folding (# of folds at 1 kg. tension MD/CD)	Tensile (pressure required to pull fibers apart, in lb. MD/CD)
100% rag high-quality printing paper	231.1	22.5	7.0	65	320/320	370/380 (rough)	37/59	19.8/24
Newsprint	33.0	3.0	4.0	13	24/36	105/115	10/29	3.0/8.0

STUDIO PAPER TESTING: The Tamarind Lithography Workshop, 1960-70

Garo Z. Antreasian

The focus of my remarks will center on paper research at the Tamarind Lithography Workshop in Los Angeles between 1960 and 1970. As a prelude, I would like to comment briefly on the use of paper for lithography in this country before that time.

It is widely known that the unique principles of lithographic printing require properties in paper that are quite different from those of papers used for other print processes. The more important of these properties are: maximum pliability, firmness of surface, and a relatively high degree of ink absorbency. We also expect the paper to be dimensionally stable, durable, and obviously pleasing to the eye, whatever that subjective term may mean. Curiously, in the 1930s and 1940s, few papers were available in this country that could satisfy those conditions for lithography, whereas for the etcher, there was a rather wide variety of exotic and rather beautiful, soft and hard finish, laid and wove papers available.

Interestingly enough, one may find some American lithographs by Bellows, Davies, Pennell, and a few others from an earlier period printed on fine papers such as Whatman, Barcham Green, Van Gelder, and Fabriano, to name a few. But, this was usually only before 1930. After that time, virtually the only papers in demand for lithography were Basingwerk Parchment and some Rives papers in light weights. BFK, which is the Rives paper we are especially familiar with today, was not even available in this country during that period. Now, I don't want to fault Basingwerk Parchment, and I don't want to fault the lighter weight Rives papers. They're perfectly fine and perfectly beautiful papers. I only want to say that the dominance of these two papers for all lithographic duties underlines the narrow confines and limited attitudes within which lithography prevailed during this uninspiring era.

We should also not overlook that, in that period, the majority of work in lithography was black-and-white work, not color printing, and most certainly not multiple color printing, because virtually none of that was practiced in the United States at the time. Consequently, these two makes of paper gave very good service for the very limited uses of the process.

The "Print Renaissance"

By the late 1950s, the demand for a greater variety of papers for all types of printmaking increased as our experience and attitudes changed during that lively period referred to as "the print renaissance." I recall vividly the impact of hundreds of color lithographs and the great variety of beautiful papers that these were printed on at the Cincinnati Museum's color lithography international exhibitions.

I also remember a rather large lithograph about that same time that had been produced by Nathan Oliveira that was printed on a heavy coarse paper larger than that which was in common use then. It turned out to be paper that we now call German Copperplate, and that paper began to be used extensively by the few American lithographers who were working with color in the 1950s. In my case, I used that paper almost exclusively between 1955 and 1960 because of its weight and size and its special quality to absorb multiple layers of ink.

To refresh my memory about the paper story at Tamarind, I went back to the archives now stored at the University of New Mexico and was astonished to find that more than 50 separate binders record in great detail the voluminous correspondence, test reports, and exhibits confined to the subject of fine papers alone. These records provide a remarkable view in depth of the paper research program and reveal especially the singular vision, passion and persistence of its founder, June Wayne.

Of equal importance, but far less well known were the efforts of Vera Freeman and George Nelson of the Andrews/Nelson/Whitehead firm whose genuine devotion to the cause of fine paper carried on much of the search of quality papers on Tamarind's behalf. Certainly Tatyana Grosman and later Ken Tyler and numerous others deserve much credit for making us all paper conscious during the decade of the 1960s. But mostly, I feel the artistic and economic climate of the time was just right for the great paper revival to happen when it did.

Four Papers

Four papers were used at Tamarind during its first year of operation, and they remained at the backbone of its activity throughout its existence. They were: Arches Cover, white and buff in 22″ × 30″; Rives BFK and German Copperplate were used for large prints; and Inomachi Nacre was reserved for the Tamarind impressions because of its unique and lustrous appearance. As Tamarind's experience grew, we were able, by

the second year, to use these papers to define desirable and undesirable paper characteristics under varying working conditions for direct impression lithography.

Thus began the first systematic appraisal of fine papers for hand printing lithographs since tests by the European lithographers such as Englemann, Hullmandel, Lorrileux, and Lemercier at the middle and early parts of the 19th century. The testing program began by establishing a basic nomenclature of paper properties and working procedures by which each variety of paper could be tested and recorded in the same way.

The report form included the following: the name, color, finish, dimension, weight, number of sheets per package, and the manufacturer. Then the description was recorded, such as the hardness of the sheet, the character of the sizing, the type of the edge, direction of the grain, type and location of watermarks, density of the sheet, caliper thickness before and after various passes through the press, the pH measure, and whether the paper was handmade, mold-made, or machine-made.

Each sheet was tested for its printing adaptability to varying drawing techniques used in lithography such as washes, solids, and the various crayon methods. Also, all papers underwent Fadeometer tests for light fastness and tests to determine ink trapping ability and tendencies to curl, pit, and waffle. Negative characteristics were reported, and the report concluded with recommended uses for the paper if such were warranted.

Once underway, the searching and testing of fine papers to meet high standards gained rapid momentum.

As the varieties of papers proliferated, their standardization did not always keep apace. The problem about standardization is that in production printing, where identical impressions are essential to the reliability of the workshop, then a standardization of paper performance becomes imperative. Predictability and performance from one lot of paper to the next was sometimes less than desirable.

One serious crisis surfaced in 1963 when newly arrived stacks of Rives appeared with a slightly glazed surface and curled edges. The sheets seemed less firm and reluctant to accept multiple printings. Representatives, from the mill, from Tamarind, and from Andrews/Nelson/Whitehead met repeatedly in New York, Los Angeles, and Paris to trace the problem.

A hand press was installed at one point in the Rives mill. Identical tests were run using our inks and using French inks. After months of exchange, the mill representatives agreed that there was indeed a change in the behavior of the paper, but they could not account for it.

One conjecture was the possibility that the world's rag supplies contained gradually increasing amounts of synthetic fibers, such as nylon and dacron. This possibility, together with greater quantitites of cotton linters that would appear later in the rag pulp, could account for the surface sheen and ink rejection that we were complaining about.

In her progress report in 1963-64 to the Ford Foundation, June Wayne wrote, "Concurrently the supply of handmade, all rag paper, is choking to a trickle. Our present supply of French paper is all we can count on. No knowing what substitute we can develop."

Despite this gloomy outlook, she redoubled her efforts by traveling to Europe again and again to search out new papers. At one meeting with mill representatives in Paris, she was told that the Rives riddle seemed to have been caused by a foreman who had run into a shortage of some factor for the furnish of the paper. To solve the problem, without interrupting production, he had altered the pulping formula but was no longer able to remember the old percentages, which also could not be reconstructed by the management.

Incredible as this may sound, we should also remember that papermaking in Europe was, even at that time, in a state of transition between the old artisanship of unwritten formula and today's much more scientific quality control. Over that year alone, the Rives mill made five samplings of paper, but in our view (that of the user) none constituted a return to the old.

New Fine Papers

Meanwhile, other old and venerable mills were reluctant to undertake speculative ventures to fabricate new fine papers. At that time, the early 1960s, their production schedules and machinery were heavily committed to producing currency papers for the fluctuating economies of South American, Asian, and emerging African nations.

Nevertheless, progress was made and accelerated testing was underway in 1965 as papers began arriving from Italy, France, England, and Germany. By that time, some 10 papers ranging from Copperplate Deluxe to De Laga Velin Narcisse were thoroughly tested and approved by Tamarind standards. It is significant and appropriate that subsequently all of these papers have found a wide and profitable market in the United States.

The culmination of one aspect of the paper search program was the publication titled *The Beauty and Longevity of an Original Print Depends Greatly on the Paper That Supports It*. This small pamphlet contains samples of pertinent technical data of the major fine papers that had been approved by Tamarind standards up to 1970. Of course, another publication was the Tamarind book on lithography of which I was a coauthor. The chapter on paper, in this publication, could not have been written without the results of countless tests that were run by a great number of Tamarind Printerfellows.

Paper testing continues at Tamarind Institute at the University of New Mexico, but not at the same feverish pace that it did earlier. I am told that an update on some recently tested papers will be published in the *Tamarind Technical Papers* in the near future.

More significantly, Tamarind laid important groundwork in the 1960s. It revived the high standard of artisanship in the service of art and deepened our knowledge and attitudes about the technical and aesthetic potential of fine papers. In my view, the current generation of papermakers in this country are in one way or another inheritors of that legacy of dedicated professionalism that sprung from those early Tamarind programs.

Garo Z. Antreasian, former technical director of Tamarind Lithography Workshop, Los Angeles (1960-61), and of Tamarind Institute in Albuquerque (1970-72), is coauthor of the Tamarind Book of Lithography: Art and Technique, *1971. He is a respected printmaker and is currently professor of art at the University of New Mexico, Albuquerque.*

CARE AND CONSERVATION OF WORKS OF ART ON PAPER

Inge-Lise Eckmann

The initial step in caring for art works is understanding the physical and chemical properties of art materials and their aging characteristics.

A thorough understanding of art materials requires considerable study, but with careful visual examination, one can identify common materials and determine some of their properties. In addition to this information, some knowledge of the agents that hasten the deterioration of the materials provides one with the tools to construct more stable art works and contribute to their preservation through proper maintenance.

Properties of Paper Supports

Paper supports are cellulose-type fibers cast in a felted structure. The longer the fibers, the more they will be interlaced, and the more resistant to damage the sheet will be. Cellulose is a hygroscopic material, and it reacts dimensionally by expanding when moisture is absorbed from the atmosphere and by contracting when moisture is lost. The degree of dimensional change varies in different papers (increasing, for example, if the fibers have been extensively beaten), but some dimensional instability should always be anticipated.

Cellulose is a chemically sound material, and pure cellulose papers, such as cotton or linen rag papers, have great longevity. The addition of noncellulose materials to a paper support can drastically reduce the stability of the paper. Lignin, which is commonly found in 19th and 20th century wood pulp papers, and acidic sizing additives present in even older papers are two common components that hasten the degradation of paper.

In the deterioration process, the long-chain glucose molecules are broken down, resulting in shortening of the fibers and a weakened structure. As well as becoming more brittle, papers tend to discolor during this process. When acidic noncellulose materials are present in a sheet that is made up of excessively beaten short fibers, the paper deteriorates quickly. Knowing that certain papers are particularly unstable should deter the artist from using them, but the low cost and ready availability of these materials has led to widespread usage.

If an artwork is especially discolored or brittle, it may be an inherently poor quality paper. In order to slow down the aging process of any paper, and especially of any poor quality paper, it is necessary to reduce the object's exposure to environmental factors that contribute to deterioration. Paper is a permeable material and therefore greatly influenced by its environment. Contact with grime, pollutants, or acidic wood pulp matting materials often result in the breakdown of even good quality paper supports.

Characteristic properties of paper to be noted when examining an art work include its thickness, flexibility, surface texture, mechanical strength and permeability. Optical properties of color, surface gloss, opacity, and the presence of sizing or coatings should also be noted. Through recognition of these factors, one can become aware of the more fragile characteristics of the paper and avoid damage.

Properties of Media

The media of a print, drawing, or painting usually consists of the coloring agent, which is a pigment or a dye cast onto an inert filler, and the binding medium. The ratio of pigment to medium may vary from a rich film with a relatively small quantity of pigment immersed in the medium, to a lean film in which the pigment is bound with little or no medium. A rich film usually has a saturated appearance and tends to have more cohesive strength than a lean film. A lean medium has a less saturated appearance due to light refraction in the air spaces between the pigment particles. The minimal amount of binder between the pigment particles gives the lean film less cohesive strength than the rich film. A pastel drawing is usually a lean film, and an oil ink print is usually a richer one, but even a traditionally rich medium can be diluted and applied as a lean film. Each object must be examined for the structure as well as the components of its media.

The structural soundness of the design layer is affected by the flexibility of the medium as well as the relative proportion of pigment to medium. If a medium is brittle or becomes brittle with age, the art work will be susceptible to mechanical damage from dimensional changes in the support or careless handling. Any apparent cracking or cleavage of the media from the paper would indicate some brittleness in the design layer.

Another factor in the strength of the media is its adhesion to the support. The design layer is bound to the support by mechanical action as well as by the media's adhesive properties, so the method of application of the media is an important factor.

The surface texture of the support and the amount of pressure employed in the application of the media affect the lamination. Charcoal drawings are traditionally executed on textured paper. As the charcoal is drawn over the surface, the particles are deposited on the raised fibers, and this mechanical action binds the charcoal to the paper. An art work with this fragile structure will require exceptional handling, as the pigments can be easily dislodged. An oil ink print will not be as friable, but the raised ink on an intaglio print makes is susceptible to surface abrasion, and any alteration in the smooth surface of a serigraph is difficult or impossible to repair.

In addition to examining the structure of the media and its adhesion to the support, the thickness and weight of the media relative to the thickness and strength of the support should be assessed. Applying a thick film to a thin or brittle support will create a weak structure.

Chemical instability in the coloring agent may lead to fading or darkening. Light levels in exhibition areas should always be controlled, but if there is any indication of fading or discoloration in an art work, its exposure to light must be reduced. Solubility of this media, which can be determined only through testing, is a property critical to the conservator.

It is often difficult to determine whether there is a sizing or coating under the media or a fixative or final coating on the surface. Any apparent layers should be examined and such things as color, thickness, evenness of application, and damages or alterations with age noted.

Agents of Deterioration

Certain factors in the environment hasten the deterioration of materials. Recognition of the physical changes associated with the degradation process can help to determine which process has occurred and the steps that can be taken to retard the process.

When an object is constructed of unstable or incompatible materials it will have an inherently poor structure and a limited longevity. Incorporating materials such as acidic wood pulp papers or fugitive dyes will create art works with what is called an inherent vice characteristic. Inherent vice can also result

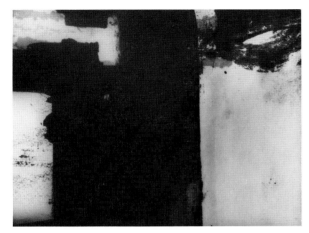

1. Detail: Franz Kline. Untitled. *Oil paint on paper. Private collection. Originally a black drawing on buff newsprint, this Kline has become multi-colored as the linseed oil has discolored the newsprint, creating orange- and brown-hued shadows to the black image.*

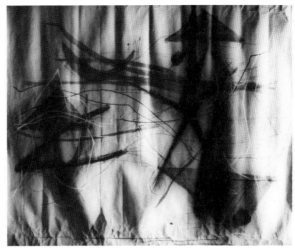

2. Adeline Kent. Drawing. *Ink and waterbase paint on kraft paper. Private collection. To avoid wrinkles and creases when rolling a work of art, one should roll it around a stiff tube of large diameter, drawing side out, always protecting it with a layer of tissue or glassine and an outer stiff tubing for mailing purposes.*

from combining materials that do not respond similarly to changes in temperature or relative humidity. Fabric and paper, for example, respond inversely to moisture absorption (paper expands and fabric shrinks).

Chemical incompatibility can occur when, for example, excessive amounts of acidic linseed oil medium is applied to paper resulting in staining and embrittlement (*Fig. 1*). A basic understanding of the properties of materials makes apparent the risks involved in handling or fabricating a structure, such as a heavy or brittle design layer on a thin or flexible support. Regardless of how well an object is constructed, there are risks that arise when it is handled, and these risks are multiplied when the support is large or has an unusual shape.

Handling Art Works

Improper and careless handling are factors in deterioration that can be avoided with care and common sense. It must be assumed that each art work is fragile and deserves full attention when handled. Art works should be handled with two clean hands and placed on a stiff support when carried.

No art work should be rolled, but if this cannot be avoided, roll it around a stiff tube of large diameter. Roll the object with the design layer facing out and a piece of tissue or glassine rolled with it to protect the surface. Although the media may crack if rolled face out, it will not develop buckled cleavage or loss from being compacted (*Fig. 2*).

The media is fragile and should never be touched. It is best to mat all art works even if they cannot be framed because this housing provides protection for the object.

When unframing art works, do not use sharp tools, and remember that old mats may be very brittle. When examining an object, never open the window mat from the inside, do not attempt to remove mounts or mat if adhered to the object, and keep potentially damaging objects such as tools or writing materials safely on another table.

3. Detail: Hall. Creation. *Oil ink woodcut on paper. Private collection. "Foxing," the brownish spots that grow on the paper when microorganisms find sufficient humidity (clearly seen in the margins), can be stopped by chemical means from further degenerating the paper.*

Light

Excessive light, particularly the ultraviolet wavelengths, catalyzes the photochemical degradation, which results in discoloration and embrittlement of paper and the fading of some media. Acidic papers and fugitive dyes are more susceptible to the effects of light than rag papers or carbon inks.

It is always desirable to avoid exposure to ultraviolet light and reduce visible light levels. The use of ultraviolet filtering sleeves over light fixtures or ultraviolet filtering plexiglass over objects is ideal. Using incandescent rather than ultraviolet, rich fluorescent lighting is advised, and art works should never, of course, be exhibited in direct sunlight.

An optimum visible light level is 10 to 15 foot candles. Foot candles can be measured with a light meter. To approximate the foot candle reading with the light meter in a camera set the meter at ASA 80 and aperture 5.6, and focus on a white card. The denominator of the resulting shutter speed will be the foot candle reading. Ten foot candles may seem like an inadequate amount of light, but if the ambient light in the exhibition area is low viewers become accustomed to lower light levels and accept them.

Humidity and Temperature

Variations in temperature and humidity are interrelated factors that affect art works in many ways. Increases in relative humidity occur when the temperature drops outdoors, in a room, or inside the microclimate of a framed art work. If the relative humidity rises to 70%, mold growth can develop on paper resulting in staining and damage to the sheet. "Foxing," the common brown spot stains, develops when microorganisms in a sufficiently humid environment act on papers (*Fig. 3*). Mold weakens objects because it feeds off of paper, sizing and gum, or glue-base media. The mold action can be stopped with fungicide, and in some cases the stains can be reduced with bleaching. But the paper remains weakened in the areas which have been affected by the mold.

Avoid creating an environment that supports mold growth by keeping the humidity constant at about 50%. Never store objects in cold environments such as basements, or exhibit them against cold exterior walls.

Paper should never be framed in direct contact with glass. Moisture readily condenses on the surface of cool glass, and if paper is in contact with glass, it will absorb this moisture. Even if the humidity in the environment is not high, objects framed against glass commonly become distorted and sometimes adhere to the glass.

When excessive moisture results in the distortion of the support, cracking and cleavage of the media can occur.

Excessive dryness, often associated with high heat, can cause shrinkage of the support and breakage if the support is restricted from shrinking. Media and paper are generally more brittle and fragile when extremely dry. As well as affecting the physical properties of art on paper, heat increases the rate of degradation of art works. Never exhibit or store objects near heat sources, and keep the temperature of the exhibition area cool.

4. Family Register. *Oil ink and writing ink on paper. Private collection. Waterstaining and insect damage such as this are largely irreversible and can be avoided by providing proper storage conditions.*

Insects

Insects that damage art works include those that feed off of glue sizing or starch paste, such as silverfish or cockroaches, and those that attack cellulose, such as termites or woodworms. Insects are a common problem in storage areas because they prefer a warm, dark environment.

General cleanliness and regular inspection of storage areas can help prevent damage from insects. If insects are present, it may be necessary to use aerosol or powdered insecticides, taking care, of course, not to expose the objects to these materials (*Fig. 4*).

Pollution

Pollutants carried in the air are an ever increasing problem. Common industrial pollutants that affect art materials include sulfur dioxide, which bleaches, discolors, and embrittles paper, and hydrogen sulfide, which darkens lead pigments. Airborne soot and dirt create stains and carry moisture.

When dust is deposited on the surface of an object, the moisture it holds can be transferred to the object. Maintaining

a clean indoor environment by frequently replacing filters on air and heating vents and by isolating each object in boxed storage or frames are the best ways to guard against the effects of pollution.

Storage

Conditions for good storage include a clean area with stable temperature and relative humidity. Works on paper should all be matted. If it is not possible to mat every item, mat those with the most fragile media. Sort unmatted works according to size and stack them with acid-free interleaving. A smooth-surfaced, nonabrasive paper such as acid-free glassine is best. Place all unframed objects in acid-free boxes or map drawers lined with acid-free paper.

Regular examination of collections in storage as well as those on exhibition is the best way to provide overall care for a collection.

Matting and Framing

Proper matting and framing provide excellent protection for a work of art on paper. The use of improper materials or techniques can, however, cause tremendous damage.

Due to paper's permeable structure and hygroscopic character, it is greatly affected by the materials with which it comes in contact. Acidic components of wood pulp mats can migrate into paper. Mat burn, the dark stain often seen just inside the window of a wood pulp, matted object is a result of this acid migration. Mounting or spot gluing of paper to a backing is undesirable because this restricts the paper's dimensional response to changes in humidity and often results in distortion and breakage of the support.

Art work should be secured into a 100% rag (cotton fiber) mat with paper hinges (*Figs. 5 & 6*). Hinges attached minimally to the top edge of the support allow for dimensional change. A four-ply thickness of rag board is best. This thickness affords support in handling, and a window that is deep enough to separate the delicate surface of the work from the glass.

The window can be adhered to the backing with linen tape, but the correct material for the hinges is Japanese tissue

5. *A long-fibered Japanese paper hinge will cause the least number of problems with buckling due to humidity.*

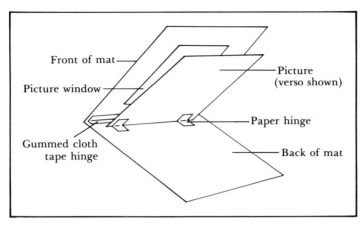

6. *Proper matting technique.*

Front of mat

Picture window

Gummed cloth tape hinge

Picture (verso shown)

Paper hinge

Back of mat

adhered with starch paste. The long-fibered Japanese papers have great strength and do not distort excessively when wet. Starch paste is a strong adhesive, which does not discolor with age, and can later be removed if moistened. Particularly undesirable materials that have commonly been used to secure paper objects into mats include acidic kraft paper tapes, dimensionally unstable glassine paper tapes, and pressure sensitive adhesive tapes or adhesives which cause staining and are often difficult or impossible to remove.

Hinges should be small. Any material adhered to the support will cause that area to respond differently to moisture changes. If the paper is heavy, use more than two hinges or expand the width of the hinges, but don't extend the hinges down into the image area. A hinging paper should be chosen that is somewhat thinner than the support paper, and the edges of the hinges should be torn rather than cut to avoid any distinct impression in the support.

Wheat or rice starch pastes are used by most conservators. There are many starch recipes, but the following is a common recipe: Measure by volume one part starch and four-and-one-half parts water. In the top of a double boiler, mix the starch with about a third of the water. Heat the remaining water to a simmer, stirring constantly. Add the hot water to the starch mixture. Cook this in the double boiler and continue stirring to avoid scorching. When the paste thickens and becomes translucent, it is ready to use.

Although cooking paste takes approximately 20 minutes, pure starch is recommended over commercial pastes, which contain unknown agents that may reduce the stability of the paste.

The paste should be brushed onto the hinge thinly and evenly, and it should not contain much water when applied. Immediately after attaching the hinge, place a small blotter and a flat weight on the area. Allow the hinge to dry for several hours before removing the weight in order to insure a good bond.

Paper should always be framed with a protective glass or plexiglass. This glazing must be separated from the support, and if a traditional window mat is not desired, rag board strip spacers can be secured behind the glass at the extreme edges (behind the frame rabbet). Plexiglass has the advantages of being lightweight, less breakable than glass, and available with ultraviolet filtering properties. It does, however, carry a static electric charge that attracts dust and other small particles, such as loosely adhered pigments. Plexiglass is, therefore, not recommended for media such as pastel.

Behind the matted object, a stiff backing board should be placed and secured with brads. A brad squeezer will set the brads without jarring the object. If the frame is not hardwood, brads can be set with pliers. Hammering is not recommended. The reverse of the framed object should then be sealed from dust and grime with an overall paper backing or paper tape around the edges.

Conservation Techniques

Each art work is made up of a unique combination of materials constructed in a specific way. After its creation, the work has been exposed to various environmental factors and has aged. Only an experienced conservator can understand the materials and changes that have occurred during aging well enough to determine the form of treatment that is applicable to an object. Improper repair can result in irreversible damage.

The conservator's goal is to preserve the original. Restoration, or repair and compensation for loss, is the second consideration. There are many treatment procedures, and the proper techniques and materials can be selected only after a thorough examination has been carried out.

A proposal of treatment and record photographs are prepared before treatment is begun. It is important that treatment be documented and that materials that are applied to treat the object are not only compatible but reversible. As the object ages and is exposed to various conditions, it may be necessary to remove repairs in order to treat the work again.

Removal of old repairs or matting materials is often the first step in conservation treatment. Aged mats or backings are usually brittle, and often the paper support is even more fragile than the mat. Hasty removal of mats or tapes is likely to abrade or tear the support. Even if the matting materials are poor quality, it is safer not to remove them yourself if they are adhered directly to the support. Their removal is a delicate mechanical procedure often requiring the use of an appropriate solvent, which will soften the adhesive without affecting the object adversely.

To reduce grime deposits the borders and reverse can be dusted with a soft brush. Careful manipulation of powdered eraser on the borders and reverse can reduce grime further, but paper is easily abraded, and particular care must be taken not to disturb the media.

Cellulose swells with the application of moisture, and water can be used to relax distortions in paper. Use of uncontaminated water, either in spray washing or immersion washing, can also reduce the concentration of discoloration products trapped in aged papers.

Applying moisture to an art work can be beneficial, but it can be disastrous. It is extremely difficult to determine exactly how the support and media will react to the application of water. The conservator will test for the water solubility of the media and will try to determine the expansion characteristics of the paper before moisture is applied. Paper, especially aged paper, is extremely fragile when wet, and handling involves great risk. If carried out properly, water washing can improve the appearance of an aged paper greatly and can even result in renewed flexibility.

In order to reduce stains such as foxing or mat burn, the object may require bleaching. This procedure can improve the appearance of an art work, but the deterioration that has occurred in the stained areas cannot be reversed. Bleaching exposes an already damaged paper to extreme chemical stress and is not recommended unless the staining is disfiguring.

Deacidification is a process in which an alkaline buffer is introduced into the paper. The application of deacidification agents (often calcium or magnesium bicarbonates) can help retard the acidification and degradation of paper.

The repair of tears or punctures in the support are carried out with starch paste and reinforced on the reverse side with small patches of lightweight Japanese paper. Voids can be filled with inserts made from a paper similar to the support or cast from a rag pulp into the form of the loss.

A support that is extremely dessicated or has numerous tears may require overall reinforcement. A backing of thin Japanese tissue is then adhered with dilute starch paste. Application of sizing may be required to add strength to a deteriorated paper.

Compensation for lost media is usually kept to a minimum. Media applied to paper will be difficult or impossible to remove in the future. Painting in areas of loss is often carried out with a lean dry media and restricted primarily to the conservator's inserts. The intention is to reduce the distraction created by the damage rather than the complete restoration of the area.

There are many other methods of treatment available to the conservator, and the intention here is only to make the artist or collector aware of the conservator's work. The degree of success that can be expected from a conservation treatment is always limited by the components and condition of the object. Although a damaged art work can often be repaired, some effects of the damage remain.

The key to the preservation of art works is to take precautionary steps to avoid damage and to slow down deterioration through proper construction.

Inge-Lise Eckmann is chief conservator of Works of Art on Paper and Contemporary Paintings for the San Francisco Museum of Modern Art.

AN AESTHETIC HISTORY OF PAPER IN PRINTS

Andrew Robison

I want to make it clear at the beginning that I have almost no professional qualifications to talk to you. I am not a professional printmaker; I have never made a print. I am not a professional papermaker; I have never made any paper. I am not a paper chemist. I am not even an art historian.

About 12 years ago, however, I just went bonkers about prints. I love to look at them in all shapes, sizes, periods, colors. That background of personal desire and collecting, I guess, is the reason I am a curator, so that now I participate in collecting with other peoples' money, and on a grander scale, which is very nice.

One thing I have tried to do in this field, as an amateur and with my background in philosophy, is to close what struck me as an outstanding intellectual gap, which was the subject of my exhibition and catalog, *Paper in Prints,* published by the National Gallery of Art, Washington, D.C., 1977. The gap I am referring to is that left between publications and exhibitions on the history of papermaking (the kind of subject Dard Hunter did so well, if not definitively) and, on the other hand, the subject we see so frequently in histories of graphic art, that is, the history of graphic techniques. When you pick up a book on the history of printmaking, for example, you typically read something like, "Woodcut was the earliest medium . . . and here is the earliest woodcut. . . . Here are the later ones. . . . Then

came engraving. . . . Here is early engraving . . . etc."

You can recognize that this contemporary approach is not the only one when you remember that the older collectors were interested in graphic art because of its images. For example, they collected certain Dürers because they showed religious subjects. When you look at old albums, they are organized according to subject, such as, *St. Jerome* through the Ages, or *The Adoration of the Magi* through the Ages.

Then, in the late 18th century, the idea arose that what was important about categorizing and surveying graphic arts was not the type of images, but the media, that is, the printmaking techniques through which these images were created. This meant we should look at the history of graphic art through the history of these techniques—engraving, then etching, and so forth.

What seemed to me to be missing, however, was a survey of graphic arts from the point of view of the third component of any printed image, its support. Any printed image has several components. First, it has a matrix, some sort of worked matrix, which involves processes, that is, techniques through which the image is created on the matrix and transferred to a support. Second, it has a subject matter. And last, it has a support, which in most cases in the West has been paper.

There did not seem to be a history of paper as used for

prints. I do not mean paper by itself, nor prints by themselves independent of the support. Rather I mean the use of that paper support by the artist in its original condition. In other words, I am talking about the aesthetic use of paper in prints and what positive or negative roles the paper plays in the printed image.

Well, that is the subject I tried to discuss in *Paper in Prints*. As a rank amateur, I am sure that there are many mistakes in what I have said, but at least it is an attempt.

What I am going to try to do in this talk is to say something about the same subject—the aesthetic uses of paper and the variety of the aesthetic uses of paper in various fine prints. I also assume I have been asked to come here from the stodgy old National Gallery to represent the old guys. So, I am going to talk mainly about the old guys, that is, the old masters, rather than the contemporary uses of paper. My focus will be on some of the origins and some of the distinctive characteristics of paper and its aesthetic use in printed images in earlier centuries.

For purposes of trying to give some organization to what I am going to say, I will divide the aesthetic uses of paper in terms of basic physical characteristics of paper—its *surface*, its *color*, and its *size*. Any piece of paper, no matter what kind of scrap nor how elegantly made, has some kind of surface, color, and size. Each of these factors can play an enormous role in the aesthetic use of the paper by a printmaker.

Surface: Texture

I focus first on the surface qualities of paper, independent of its being printed. Before it is printed, the paper obviously has a certain texture. There is the obvious difference between laid texture and wove texture, and you are familiar with the kind of mechanism that produces these different textures.

What I want to focus on, then, is not the general distinction between laid and wove textures, but further distinctions below that level. This brings me to the reason why I am showing you objects today rather than slides or other reproductions of those objects. The problem is to show qualities so subtle they just are not captured in normal reproductions but require face-to-face awareness of the objects themselves. During the talk, I will hold up the objects and describe these qualities. You

1. Anonymous (South German). Adoration of the Magi. 1400-1410. Woodcut. Courtesy National Gallery of Art, Washington, D.C., Rosenwald Collection. One of the earliest European prints on paper.

are welcome to come up afterwards and see these objects at close hand so you can test my descriptions.

Laid or wove is a very important distinction, but going beyond that, one can also say that in the history of graphic arts, there are certain distinctive textures that appear at different times. Naturally, the texture arises because of the papermaking materials—the substances that are used for the paper, the way the fibers are macerated, the way they are matted or formed into sheets of paper, and the way that paper is treated after the sheet is formed, whether it is finished in some way or sized or polished.

One does find, for example, a very distinctive kind of texture in the very earliest paper used for printed images in the West, that is, in the 15th century.

Here is one example of a print in pristine condition from the earliest printed book with images made after a few rare editions in the 1460s. It was printed about 1471-72 by a man named Gunther Zainer in Augsburg. It shows a nice little print of St. Augustine trying to convert a couple of unbelievers. Now, when you take this paper and get it into the proper raking light, you see that is has a very distinctive texture. Not only do the laid and chain lines show up quite clearly, but also you can notice numerous tiny indentations, little curved lines, some light straight lines, but mostly little curved lines. They look like the remainders of hairs.

Indeed, if Dard Hunter is to be believed, the reason for this lies in the way paper was manufactured in the 15th century, which illustrates another relationship between art and technology. At that time, after the paper was pressed between felts, it was not finished further, or finished only minimally. So, what we see in these earliest prints is the remainder of the hairs from the surface of the felts, indented with pressure into the wet paper, which left the tiny indentations.

Note, however, that we see them only when the print is in pristine condition. If you take this Zainer print and wash it and iron it, then the indentations of the hairs disappear right away. However, the telltale hairs and indentations can be seen again and again if you look through, for example, books where the sheets have been bound together and left alone, or in the rare pristine examples of separate prints.

One finds this distinctive texture, not only in woodcuts like Zainer's, but even in engraving where finer paper is involved, such as this Dürer engraving. This appears to be one distinctive characteristic of texture in paper form the 15th, 16th, and 17th centuries.

Other distinctive features of textures can be found in other periods. For example, the kind of paper manufactured in the 18th century in North Italy (presumably around Bologna, perhaps as far south as Fabriano, but primarily around Venice rather than farther south in Rome) also has a very distinctive texture. It has very broad ribbing lines.

When, for example, Tiepolo is closely examined, you can see that as opposed to even the woodcut paper, which was a rougher, grosser 15th century paper, the Tiepolo print has laid lines that are very broad, with a rough surface—rough in the sense that the laid wires have left quite clear, widely spaced indentations. Again, this is a distinctive texture that occurs at a certain period.

It is possible to elaborate such examples much further. Take the kind of pulpier, chalkier paper that appears in the early 19th century with early wove paper, and so forth.

Surface: Receptivity

Another crucial quality of paper related to its aesthetic use by a printmaker is paper's receptivity, or how well it receives the ink. One revealing example of the problem of receptivity of paper is offered by Thomas Bewick in the fine wood engravings that he created just at the end of the 18th century and the beginning of the 19th century. Bewick printed his first wood engravings on laid paper. This proved unsatisfactory because the work on the wood was so fine that the ridges of the laid paper prevented the paper from contacting the block evenly and absorbing the fine lines regularly. So he switched to wove paper.

By 1790 Bewick was printing on wove. But even that was not very satisfactory initially. An example is in this first edition of Bewick's book on British birds. In his first edition in 1797, Bewick had already tried to find fine wove paper with a very even surface. But, unfortunately, the technology of the time and the paper that he had access to was still much too gross for his printmaking technique.

Normally, one might say that since his first edition is the earliest impression published (as opposed to a proof), it must be the best impression. Not at all. It's a terrible impression. If you look at it closely, you can see the black lines are continually interrupted by the variable absorption of the differing levels of the pebbly surface of the early wove paper. Thus, the unsuccessful printing of his engravings. Since the black lines are interrupted and the white lines are interrupted, you get a kind of splotchy appearance, as if you saw the image through a snowstorm.

In the next 30 years, the development of paper technology was sufficient to provide Bewick with marvelous paper. With one exception, Bewick had alread died by the time wove paper of sufficient smoothness was commercially produced and available for him to print his wood engravings. The one exception was the special proofs that Bewick took on a kind of yellow China paper, which he took out of the inside of tea chests. This paper was very smooth, very even, and allowed him to get beautiful proofs.

Indeed, the best impressions of Bewick's wood engravings that I have seen myself were made very recently. I have an impression of the same vignette from Bewick's 1797 book on British birds that was taken in 1972. It is also on wove paper, with a strong but soft, nice surface quality, similar to Japanese paper. It is called Troya paper. The impression was printed by Hunter Middleton of Chicago. You might think, "Here we are in 1972, almost 200 years after the wood engraving was created in 1797. What kind of impression is this for the normal connoisseur? This is a late, late restrike, not even worth looking at."

If you compare, however, the late impression side by side with the early impression, you can see that everything about the image that Bewick wanted is shown for the first time. For instance, you can see a runaway cart pulled by a horse with some boys in it, and there is a house in the background. Bewick has engraved on the horse striations of parallel white lines with a graver. In the early impression, this simply looks like a way of making the horse appear dappled gray rather than solid black. However, in the modern impression on this extremely smooth wove paper, we can see that Bewick had actually engraved the gleam of sweat on the side of the horse.

Another example from the same vignette can be seen in the white concentric circles engraved around the inside of the hub of the wheel. In the early impression, the circles make the wheel look kind of gray. But in the modern impression, where each of the black lines prints continuously and each of the white lines prints continuously, uninterrupted by the variable ridges of the paper, we can see what is really going on here. Bewick has shown us that peculiar visual effect of a wheel as it turns rapidly. One sees the moving spokes giving the impression of concentric circles from the hub out to the rim of the wheel. The distances, the recession of the planes in depth, are also affected.

You may remember Bewick's famous trick of shaving the edge of the block. In the early impression, there are trees in the background. But since the printer had to use such heavy inking and pressure to make all the work contact the paper, you get a very heavy impression and a kind of silhouette effect of the trees. In the later impression, at the edges of the trees at the very edge of the vignette, you can see that Bewick shaved off the surface of the block so that when the block is printed well on proper paper, progressively less of the edge of the block prints on the paper. As you look higher and higher in the body of the tree, the bushy leaves appear fainter and fainter so that you get a kind of recession and the tree becomes a round tree rather than a two-dimensional silhouette.

Of course, the artistic success of prints normally depends on marvelous press work, the make-ready, and the ink. But in Bewick's case, you can really see that the artistic success, the conveyance of what he wanted to show in the image, depends largely on the sophistication of the type of paper that was available. Sad only that it was 200 years after he first published it that one finally gets impressions of what he had in the block.

The problem of paper being too gross in its surface qualities, its texture, to receive the ink well to show what is really in the plate or block is a problem found in many printmakers' work, not only Bewick's. He is just a particularly clear example. There is another way that variations in receptivity can be approached, not as a problem, but as an opportunity.

Rembrandt was the printmaker who first capitalized on the variable receptivities of different kinds of supports. As one would expect, printmakers were aware of this issue already at the beginning of printing. Dürer woodcuts, for example, are on a much heavier, rougher paper than Dürer engravings. Consistently, the engravings are on finer, smoother textured paper than the woodcuts. Clearly, the artist knew what he was doing and knew that for the finer technique of engraving, one required a more sensitive, finer surface than for the rougher technique of woodcut. But that is a simple distinction that one clearly sees right away if one tries to print those two media. What Rembrandt did was to go much beyond that beginning.

Rembrandt and "Japan"

Between 1640 and 1645, Rembrandt had available to him not only European white paper, which was the standard kind of paper that he used, but also some imported paper from the Orient, what print curators and collectors normally refer to as "Japan," which apparently did come from Japan. The normal Japan that Rembrandt used has a golden color, a very smooth surface, almost a silky, smooth surface. Contrary to what one might normally expect, this kind of Japan is much less absorbent than his European paper.

Some of the earliest impressions of his prints that I know of on Japan show that it created quite a problem for the following reasons. Japan has marvelous qualities for his drypoints. The etching lines in open areas and the drypoint lines in open areas are held right on the surface of the paper so the lines are crisp and sharp. The lines are right up front. Likewise, with drypoint, any burr on the edge of the drypoint is not absorbed into Japan paper as it is in European paper, but it is held right on the surface. If you see a wonderful Rembrandt impression

2. *Rembrandt van Rijn.* Nude Seated on a Bench with a Pillow. *1658. Etching. Courtesy National Gallery of Art, Washington, D.C., Pepita Milmore Fund and gift of Hans W. Weigert. A crisp impression on Japan paper.*

3. *Rembrandt van Rijn.* Nude Seated on a Bench with a Pillow. *1658. Courtesy National Gallery of Art, Washington, D.C., gift of W.G. Russell Allen. The same etching in the same state printed at the same time, but on vellum.*

on Japan, you think you are directly in touch with the artist because the ink is right there.

This does cause certain problems, though. It gives marvelous impressions of the openly worked prints. But when one looks at Rembrandts where he portrays forms with dense cross-hatching, then one sees that in these early impressions in the mid-1640s, the first impressions he took on Japan created an artistic problem in reading the image. If you imagine that Japan holds the ink right on the surface and holds the lines of the ink distinct instead of letting them absorb slightly into the paper and bleed out, then what happens is not the same as in tight cross-hatching on European paper, where the ink bleeds slightly into the paper and fuses the areas of the tightest cross-hatching so you get a more solid black. Instead, on the normal Japan Rembrandt used, where the ink is held up on the surface and distinct, the cross-hatching is held as distinct lines so that one reads it more as a pattern of two-dimensional lines, rather than a fused area of darkness.

What happens in the case of Japan paper is that when Rembrandt is trying to portray depth by using tighter cross-hatching in certain areas, slightly varying the density to show a recession in depth of, say, a shadowed wall, then one reads the area as a flat pattern of lines. By contrast, he obtains a smooth progression in depth when it is printed on European paper, where the ink fuses slightly and gets darker gradually between the cross-hatched lines.

Typical of one of the great things about Rembrandt, in my opinion, is that having seen this problem, he turned it on its head. Instead of saying "Well, so much for Japan. That means this stuff is not good to print on," he used Japan after that to print those kinds of impressions which were supposed to be flooded with light, where the light reaches even into the cross-hatching. The light can go right into the cross-hatching because these lines are held distinct, crisp, and clear.

A similar kind of problem occurred when Rembrandt started printing on vellum. Vellum was the least absorbent of the surfaces he used since the vellum surface was very hard and repelled the ink; in a way the ink was smashed out of the lines and across the surface. The result is a kind of broad, fuzzy mark instead of a crisp line. Again, instead of giving up on vellum, Rembrandt turned the problem upside-down and used vellum to print fuzzy images. That is, he heightened this surface bleeding, as opposed to absorption bleeding. He heightened the bleeding of ink out over the surface of the vellum by using tones or films of ink on the copper plate, which he manipulated in certain ways to highlight areas or to darken areas. He then printed those impressions on vellum, which in effect smashed out the ink and made those areas of tone even darker and richer. Again, this is an aesthetic effect that the printmaker achieved because of the surface qualities in terms of the receptivity of the support.

Bewick's Problem as Degas' Opportunity

When you come to the modern period, by the late 19th century, printmakers, like painters and artists in general, wanted to show the presence of their materials and techniques to manifest the use of graphic techniques, not only to create the image, but also to show the techniques as self-evidently as possible. The standard example is always German Expressionist woodcuts. They are so strong and show the grain and gouging and so forth that you would never take one for anything but a woodcut.

This manifest use of techniques began already in the 19th century. One can see an example of the issue of receptivity of paper surface in some Degas lithographs, particularly the nudes coming out of the bath, drying themselves with a towel (*Fig. 4*). In certain cases, early proofs of these are consistently printed on the "wrong" side of the paper. Any piece of laid paper has a right side and a wrong side, that is, neither side is completely smooth, but one side is much smoother than the other, the side that lays on top of the wires of the mold.

As I said, certain early proofs of Degas lithographs are consistently printed on the wrong side so that when you look at the image you can not only see the strokes of the artist's brush or crayon to show the form of the woman drying herself, but, in addition, you can read all the lines of the paper, including the watermark. In one impression, you can see a big watermark right in the center of the woman's body!

Naturally, everything I am saying is more or less hypothetical because it is just a case of trying to put together bits and pieces from here and there to make sense out of them in

4. Edgar Degas. Nude Bathing. *Circa 1890. Courtesy National Gallery of Art, Washington, D.C., Rosenwald Collection. A lithograph printed on the "wrong" (rougher) side of laid paper, allowing the evident texture of the paper to interact with the forms of the image.*

terms of the history of technique and the history of paper. Here the question is whether Degas was doing in these proofs the same kind of thing as van Gogh with his clear and rough dabs of paint.

Again, with van Gogh you have the work of art not only creating an image and showing the image but also evidencing the means through which the image was created. A dab of paint not only shows you a bit of color, which creates the rose on the cheek or something like that, but it stays as a big dab of paint that is quite evident on the surface of the canvas. Now, the Degas shows an interesting trick in printmaking to achieve the same kind of effect that was being done in painting at the same time. Of course, the story goes on into the 20th century with more familiar examples, such as the blotting paper the German Expressionists used.

Surface: Alterations

Besides considering the surface of the paper before it is printed, its texture as it is printed, its receptivity, we should also look at the role that the surface has after printing in terms of whatever alterations in the surface that may have been made in the printing itself. It is obvious that the original surface qualities of paper in a normal woodcut are going to be much more evident after the printing than in a normal engraving or lithograph. In an early woodcut, for instance, when you look at a fine impression in good light, you find in the image an interaction between the lines and flats of the image that have been pressed into the paper and the more or less untouched white paper in between. This paper still manifests all those characteristics of texture that were true of the paper before printing—not only the laid lines and chain lines but also the indentations of the felt hairs, etc.

As opposed to that, an engraving or lithograph, where there is a plate outside the lines of the image, has a paper surface that is altered by the pressure of the plate and is polished, to some extent. The effect with a Dürer engraving, for instance, is that the paper surface has acquired a kind of sheen. In a simple-minded way, we might say it acquires a sheen that is very appropriate for the finer linear and tonal technique of engraving. Just as Dürer's engravings used very fine lines, created fine textures and so forth, his woodcuts have lines that are rougher and grosser, more powerful. In other words, the alteration in the surface of the paper through printing the different techniques is appropriate to the technique.

The specific and intentional alteration of paper surface to give different levels of that surface, that is, embossing, had already begun in the 15th century. There are remarkable examples of 15th century prints printed without any color whatsoever, just embossing, white on white. The question of when intentional embossing in woodcuts with ink began is a very difficult one to answer. Certainly by the 18th century. In the work of artists like John Baptist Jackson, you find the leaves of trees molded in three dimensions as well as printed with some ink or color to delineate them in two. Another example of the early revival of embossing were the prints by Pierre Roche about 1890-1900. They have a very mild bit of color on them, but primarily the image is created through the third dimension, through the embossing.

I am sure you are familiar with the story about the German Expressionists. They capitalized on the soft qualities of the blotting paper they used in printing by banging the woodcut blocks right into the paper and creating large convex ridges of paper that pop forward.

Later, in the 1930s, Hayter used this same principle, that when white paper comes forward in three dimensions, it comes forward optically as well as physically. The level that the white paper is brought forward makes an enormous difference because of the way the light falls on it and the shadows that fall behind. The result is that if the artist reserves some area of white within an inked area, even in an abstract image, he can do the same print but cut out the plate where that white area is so that the paper is embossed forward and the white just screams at you. It pops right out in addition to creating a different physical dimension or level in the print.

Other contemporary printmakers who capitalized on this principle are Johns, Albers, and Rayo, where different levels of embossing create different spaces within the same print.

Color: White, Whiter, Whitest

Beyond qualities of surface, any piece of paper also has qualities of color. Even white paper is not simply white, one color, but some definite shade of white. And the particular shade of white may be crucial in portraying the image, as it is certainly crucial to the issue of authentic condition in dealing with old master prints. There are shades of white typical of the paper in certain periods of graphic art which arise, again, because of the specific techniques of manufacturing the paper. Paper typically has a buff shade of white through the period from the earliest European prints on paper until the 18th century. At that time, one frequently notices that the paper becomes brighter. If you compare, for example, a rather typical Tiepolo with a rather typical early woodcut, you can easily see even from a distance that the Tiepolo is much brighter. It is still not dead white or stark white—it still has a certain cream to it— but it is much brighter.

Again, the reason seems to be one of the changes in technology Schlosser refers to in his chapter, namely the invention of the Hollander beater in the 17th century and its gradual adoption throughout Europe. In earlier centuries, fibers for paper were macerated from rags by pounding them in a stamper, which was not a very efficient way of breaking them down. In order to enhance this process, the rags used would be first fermented in piles, like compost heaps, to break them down chemically through rotting before breaking them physically through pounding. But in this process of preliminary fermentation, the rags acquired a yellow color, even though they might be white initially. Even after washing them, these rags would retain some of this yellow or brown color, giving the buff shades of white in the finished paper.

On the other hand, with the Hollander beater, the new mechanism for macerating the rag fibers with knives instead of a solid blunt stamper was so violent that you did not necessarily have to ferment the rags before placing them in the Hollander. The beater would take care of breaking down the fibers, eliminating the need to yellow them through fermentation.

Likewise, according to Schlosser, the 18th century is the same period when one begins to find a more widespread and greater quantitative introduction of cotton into the fiber used for papermaking. Cotton is a softer fiber than flax, and so it does not need the kind of violent treatment in order to break it down and macerate the fibers. So, again, one does not need to go through the process of fermentation.

What both of these developments mean is that you can get a whiter paper. It's the brighter white that then provides the printmakers, especially the printmakers of Venice, with the paper they use to portray that brilliant and diffuse light that you see in Venice. When you go to Venice, with its clean but humid air, the light is so bright and yet diffuse. It is just what Canaletto shows in his etchings. A brilliant but diffuse light— a kind of bright ivory or cream color paper—very elegant. The same is seen in Tiepolo. The color of this paper in pristine condition is so much more elegant than the buff paper of earlier centuries.

As for the relation of cause and effect, it is sort of a chicken-and-egg problem. I mean, were the printmakers going to their suppliers and publishers and saying, "Get me a whiter paper"? Or was it the reverse, that the technique of the Hollander having been introduced, whiter paper was already available and therefore more attractive to the artist to create the same kind of image in prints that he was able to create in a painting? I have no answer to that, but, clearly, one does see the coordination between the aesthetic use of paint on canvas and the aesthetic use of ink on paper changing at just this moment when the technology of paper manufacturing made it possible for it to change in printmaking.

As for modern paper, I do not have to talk about it because you know it. You can get modern paper in all colors. However, there are some distinctive whites—the sort of dead white, the stark white, the snow white, and especially those whites with a slightly greyish or bluish overtones.

If you look through the boxes in a printroom, you can see those whites began to appear just at the beginning of the 19th century. Not surprisingly, methods of chemical bleaching of papermaking components was introduced at the same time— at the end of the 18th century. In earlier prints, one does not find that sort of stark or dead white. Indeed, this is one clue whether your print is authentic. Here is a Lucas van Leyden that has been terribly bleached through the use of some sort of chemical bleach, presumably a chlorine or peroxide bleach that has turned the paper a sort of grey-white. There is also a degree of blue-grey tinge to the paper, which you will not find in any old master print in its original condition.

If you compare Lucas with Dürer—both early impressions —you will see that the wonderful unifying buff color of the paper in the Dürer becomes a kind of stark gray-white in the Lucas. It is like the change from natural light to artificial or fluorescent light. You look at the Lucas, and there was no light in the early 16th century that would make things look like that. The Dürer is the way things would have looked.

The specific color of white can be extremely important to the achievement of the printmaker's goal. Claude Lorrain's etchings, as with his painting, look best when they have this golden buff color, which appears to be a common color of the paper that he had available. When you look at a Claude Lorrain painting, you can see his interest in the effect of sunrise and sunset in the campagna around Rome—a slightly golden color in the light. Similarly, his prints frequently have a kind of golden color when the prints are not altered by later dying or bleaching or discoloration. The light spreads throughout the image and unifies it in a very harmonious fashion.

Colored Papers

Colored papers, other than white, have fascinated printmakers since the introduction of printmaking into Europe in the 15th century. Papers were colored and used for printmaking in three basic ways. The simplest was also the earliest method. In the earliest European drawings on paper in the 14th century, paper was colored by taking a white piece of paper and washing color over it. The result is called "prepared paper." The image is drawn or printed on top of that.

Although drawing on prepared paper had been popular since the 14th century, printmaking on prepared paper was not. Printing on this paper posed serious technical problems when it was first used. From the 15th century, you very rarely find any prints on prepared paper that have survived in good condition because the color flakes off and it brings off the lines of ink with it. In the 17th century, however, there was one major printmaker, Hercules Seghers, who printed on prepared paper exceedingly well and with great success.

Often, printmaking on prepared paper was used to make decorative prints. For instance, here is a little bookcover that is printed in gold on a paper that has been washed with a lavender color so that the color is left with the striations of the brush. The striations of the washing of the color on the paper interact with the gold printing. In this case, it seems quite successful.

In the other two ways of producing colored paper besides prepared paper, color is put into the papermaking components. These two methods result in colored paper where the paper is colored throughout and not just on the surface. The first technique involves making paper from colored or dyed rags. You simply take the components that are already colored and make your paper from them.

When I first considered this issue, I was puzzled with the question of why all old master prints and drawings on colored papers occur only on blue-dyed paper, rather than on pink-dyed or yellow-dyed papers? The reason relates to the chemistry and technology of papermaking. Indigo was the most tenacious dye in general use from the 13th through the 17th centuries. In that period, paper was made from rags that were fermented and then macerated. Only indigo-dyed rags could withstand that violent process of preparation and retain the color in the fibers in the pulp used in making the paper the color from indigo blue.

If you look at the paper of early blue-paper prints carefully, you will notice that the paper has a color texture to it in addition to the surface and physical textures. That is, individual fibers or clumps of fibers are more or less dense in their color because the paper is made from macerated, colored rags where the fibers from different rags show and retain different amounts of the dye.

However, in the third main way of making colored paper, from the end of the 18th century, papers were colored by adding dye in the pulp. In such paper the color is quite even, an even and originally, a mild tone. Paper produced in this manner creates quite a striking difference in the finished product from the colored paper from dyed rags.

Printmakers have, of course, capitalized on the different kinds of colored papers. For example, in the National Gallery we have a very interesting Pissaro lithograph. It is on a kind of paper that I have not seen elsewhere, though I do not know whether Pissaro ordered it special. What Pissaro has done is to take a paper made from rags, which are both blue and red. Upon careful investigation of this paper, you can see blue clumps of fibers and red clumps of fibers very clearly. But if you step back about five feet from the print, the two mingled colors merge optically and you get a wonderful purple.

Of course, this is like some of Pissaro's paintings, where you can see little dots of single colors when you are close but when you step back, the impressionist effect takes place so that the colors harmonize and blend. Again, the technique of manufacturing the paper enabled him to duplicate in his printmaking the kind of aesthetic effect which he was clearly attempting to achieve in his painting.

Paper Size

In the the history of graphic art, size has been a much less important feature of paper for printmakers than surface and color. It has been, however, quite significant for print collectors who wanted their prints to have margins of certain sizes. Originally, collectors wanted the margins to be very small, so most old master prints are clipped to within a couple of millimeters of the platemark. Later, collectors wanted bigger margins, so prints with very large margins became popular.

The results were often very strange. For example, here is a Morin print—a 17th century French print—that was originally cut down to the borderline of the image so it could be mounted in an album like other prints of its kind. Then, in the late 19th or early 20th century, when the mania had grown for "appropriately"-sized margins, some very clever restorer took this print and added false margins around the edges. It is a very neat example since the restorer also added a new platemark within the false margin.

Thus, if you look at the Morin print, you can see everything: the margin, the platemark, and the platemark is exactly the same measurement as the size of the plate as given in standard catalogues raisonnes. I imagine this poor restorer would have had to have copper plates of every possible dimension to create appropriate sizes of new platemarks on his new margins. At any rate, this Morin is an interesting example of the foibles of collectors, like myself, who decide that margins have got to be a certain size.

As far as artists are concerned, the first really serious interest in margins began probably in the early 19th century when one finds lithographs for the first time that were bled off the edges of a sheet of paper. By the late 1870s, you find Whistler arguing philosophically about whether a print should or should not have margins and, if so, how big the margins should be.

This issue is obviously an important consideration in contemporary printmaking. A good example can be seen in one of Helen Frankenthaler's prints where the paper is left blank for a good 12 inches below the image while the image itself is printed right to the edge of the paper in the upper three-quarters. That lower foot of blank paper hardly counts as margin anymore. That is blank paper that has been fully incorporated into the image.

In earlier days, one might cut the margin off and leave the print as just the image, or not cut the margin off and leave a lot of blank edges, though with the same image in the center. However, with prints like the Frankenthaler, or Motherwells,

5. Anonymous (South German). Christ's Entry into Jerusalem. *1440-1450. Courtesy National Gallery of Art, Washington, D.C., Rosenwald Collection. A clear example of the texture of early laid paper. The woodcut shows the earliest printed figurative border, implying a conscious use of the size of paper in the sense of the otherwise blank margins beyond the primary printed image.*

or other open-field prints, the exact size of the blank paper at the edge is fully integrated into the image and must remain its original size in order to achieve the intended aesthetic effect.

Andrew Robison has been curator and head of the Department of Prints and Drawings at the National Gallery of Art, Washington, D.C., since 1974. He is also president of the Print Council of America.

ARTISTS' BOOKS

Judith Hoffberg

1. El Lissitzsky. Parolibere Futuriste.

Artists have been making books for centuries. In fact, the book was the first multiple, thanks to Johann Gutenberg in the 15th century. The revolution began then and has never stopped. Instead, it takes on new energy with each new generation and changes and flows with new ideas, new concepts, and new intentions.

As an art librarian, I have always been aware of the increased use of the book as a medium by artists, starting with the calligraphers and illuminators of medieval times, who were, in fact, artists, to the so-called "typographic revolution," culminating in the works of Marinetti, El Lissitzsky, and Moholy-Nagy, to name a few 20th century experimenters. My focus in this chapter will be on the more recent artists' books, especially those produced in the last 15 to 20 years.

In the early 20th century, the manifesto-waving Futurists, Constructivists, Dadaists, Surrealists, and others promulgated an aesthetics they applied to all phases of the fine arts that implied that the boundaries between art forms were super-fluous. In fact, we find people like El Lissitzsky, an intense and exceptional graphic designer, making books as manifestos, which still hold as visual-verbal statements today (*Fig. 1*).

These revolutionaries found that publishing themselves and each other was as much a provocative way of disseminating their views as writing poems, making pictures, or even holding mass demonstrations in the streets. All these "alternative media" were born in the beginning of our century, long before the turbulent 1960s!

After World War II, some artists surfaced who developed entirely new definitions of books and printing. Dieter Roth (whose name is also spelled Diter Rot), for instance, refused to discriminate between page and picture plane. A whole group of European artists made books in the 1950s, sometimes the precious *livres d'artiste*. These books were limited, one-of-a-kind collectibles, which were assiduously acquired by the wealthy, astute, and intelligent men and women who knew the importance, beauty, and care that reflected the artist's talents in book form.

In America, the European tradition blended with new aesthetic aims, which mixed the conscious emulation of the Dadaists and Surrealists with a united rebellion against the materialism of postwar American society. Artists joined together to perform and create the first "Happenings." They

Photos: Judith von Euer

published broadsides and spread the word of their activities in newsletters, which were examples of good graphics, fine typography, and care in the printed word. The notebooks of some of these artists (such as Kaprow, Oldenburg, and Dine) were the beginning of the new renaissance of artists' books in this country.

The Fluxus movement emerged during this period with the work of John Cage, who absorbed European art attitudes tempered with Oriental philosophy. His "Happenings," which incorporated music, performance, and dance, helped lead to the first publications and objects that became known as artists' bookworks. The Fluxus newspaper and the printed editions issued by George Maciunas as well as the more formal books and pamphlets of Dick Higgins's Something Else Press appeared between 1961 and 1968 (*Fig. 2*).

By the late 1960s, the Fluxus artists were joined by a much larger group of artists who also were producing artists' books and artists' publications. It is important to trace the factors that led to this renaissance since its impact is felt to the present day.

2. Publications from Dick Higgins's "Something Else Press".

New Printing Technology

Before 1960 the world of books was limited and expensive, and hot type and letterpress were costly. In the early 1960s, however, technological advances in the printing industry opened up a new world to artists. After years of research and development Xerox made available a machine that allowed artists as well as everyone else to copy printed works for dissemination in endless numbers. In addition, the photo-offset printing industry sprang up, which meant people then had a means of quickly printing up broadsides, books, posters, and works of art.

As a consequence, the Xerox machine and the offset press became tools of immediacy, accessibility, and convenience to the artist. Many artists had a need to communicate directly, free of aesthetic and political censorship, and many had ideas that needed a life of their own. So, to avoid the museum and gallery system, they used the mailbox as their means of distribution.

This postal distribution system makes books accessible, expendable, disposable, and inexpensive. Artists found they could send a book to friends in a mailing bag as a gift. If that worked, the artist could send it to a larger audience. One of the beautiful things about an artist's book, besides its provocative exploration of the printing medium, is that the book can be closed and opened weeks later, days later, even years later. A painting or sculpture, on the other hand, is not immediately accessible but must be located at specific places, such as museums and galleries, and viewed during specified hours. Often, too, it is inaccessible altogether, unless you can afford to buy it.

The rapid expansion of the new printing technology — photocopying and photocomposition, the computerized printing process, quick binding, and the availability of the

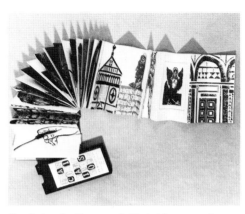

5. *Antonio Frasconi.* Kaleidoscope in Woodcuts. *1968. Published by the Republic of Uruguay to honor Frasconi as its representative artist in the 34th Venice Biennale.*

3. & 4. Ed Ruscha books. Heavy Industry Publications.

typewriter and camera—meant artists could self-publish whenever they please. A few pioneers—Seth Sigelaub, Dick Higgins, Dieter Roth, and Ed Ruscha—explored the new technology of the 1960s with great skill.

Sigelaub published the *Xerox Book,* which was an exploration of the Xerox medium by Carl Andre, Robert Barry, Douglas Huebler, Joseph Kosuth, Sol LeWitt, Robert Morris, and Lawrence Weiner. Published in 1968, the *Xerox Book* was the beginning of a continuing exploration by artists of the copying machine. Color Xerox machines are also humming with the activity of artists, as evidenced by the recent "Artworks & Bookworks" exhibition at the Los Angeles Institute of Contemporary Art.

Dieter Roth explored the phenomena of cut-up Icelandic newspapers and included it in his first book in 1957, called volume 5 of a set of 20. Similarly, Ed Ruscha, who had been trained as a graphic designer, began to make books in the early 1960s with his camera and that of his friends' to objectify his environment. Ruscha became the father of many of the developments of the 1970s not only because he produced his books well, but also because he became his own distribution system.

Distributing Self-Published Books

Wherever you go in the United States, Canada, England, France, Holland, and Germany, you can find Ruscha's books for sale. The price of these unlimited editions has risen only slightly since they were first published in the early 1960s. His library of 13 books identifies apartment houses, swimming pools, parking lots, and palm trees in Los Angeles, as well as every building on Sunset Strip (*Fig. 3 and 4*).

The biggest problem with these books does not lie in conception and production but rather in distribution. People usually learn of the existence of self-published books through word of mouth or sometimes by a notice in the mail. They are rarely available in bookstores or other commercial enterprises. As a result, many art historians and librarians have not understood this phenomenon of self-published artists' books. Commercial publishers, too, very rarely gamble on an artist's book. It is only about once every five years that a trade publisher will work with an artist from conception to execution and then sell the book in bookstores that are outlets for the publisher's more commercial publications.

6. *Eugenio Carmi & Cathy Berberian.* Stripsody.

7. *Don Celender.* Holy Holy Art Cards *and* Artball Playing Cards.

8. *Warja Honneger Lavater.* Little Red Ridinghood.

One example, however, was Antonio Frasconi's *Kaleidoscope in Woodcuts,* which was a mass-produced artist's book that had been supervised during its production by the artist and distributed throughout the world. This was one of the rare occasions when the artist's intentions were not compromised, and it was published to honor the selection of Frasconi by the Republic of Uruguay as its artist in the 34th Biennale in Venice in 1968 (*Fig. 5*).

Redefining "Book"

Some artists' books are games, some are toys, and some are objects as much as books. The traditional definition of "book" has clearly been reworked by these publications. *Stripsody* by Eugenio Carmi, for instance, not only represents onomatopoeia in which the words become a visual-verbal medium in glorious color on each page, but it also has a record inserted in a pocket in its cover that includes these sounds as verbal-sound poetry, which is recited by Cathy Berberian, the artist's wife (*Fig. 6*). The book is truly a means of intermedia.

Another example is *Holy Holy Art Cards* by Don Celender, an artist with a sense of humor (*Fig. 7*). He used the holy saint's card to satirize Harry Abrams, John Coplans, Calder, Marisol, Johns, and Warhol, to name a few. His *Artball Playing Cards* put artists' faces on baseball and football playing cards, and the effect is truly hilarious.

It should be mentioned that librarians usually put these books in special collections, away from the public, locked up in cages, to be seen only by the rare researchers, or sometimes under glass in an exhibit. This is quite different from the artist's intention to make these visual ideas accessible to anyone.

One of the oldest artists currently making books is Warja Honneger Lavater, who has been producing books for more than 20 years. Now more than 90 years old and living in Zurich, this joyful, delightful woman has explored the world of visual symbols to retell the stories of *Cinderella, William Tell, Little Red Riding Hood,* and many more. Her fold-out books delight the eye and are done with great skill (*Fig. 8*). They are books without words but appreciated by readers of all ages. Some of them are published by Adrien Maeght in Paris.

9. Long-established artists created their own books in the 1960s.

In the early 1970s, Keith Godard made books among which is *Sounds,* a book made of different papers. When you flip the pages of *Sounds,* you create sound poetry. Under glass or in a rare book collection, *Sounds* would be useless since the tactile experience of the book is part of what the artist intended to share with the "user" of the book.

Godard's *Glue* actually has two daubs of glue on the cover and explores nine kinds of glue in large illustrations, with many pages glued together in different ways. His *Itself* is a book of different colored pages with marvelous geometric forms that are cut out to tell in a few words a tale, which is a true, inner-directed philosophical story meant to touch all "readers."

Some very famous artists have also made books. Some did them as experiments, while others produced them because they feel what they want to say must be said in the medium of a book.

Claes Oldenburg, who has been making books since the days of Something Else Press in the 1960s, has recently produced *Notes in Hand.* David Hockney's reworking of *Grimm's Fairy Tales* has become a best-seller, having sold more than 60,000 copies throughout the world. Another example is Andy Warhol's so-called autobiography.

A number of other artists, such as Sam Francis, Bruce Nauman, and Nicki de St. Phalle, have produced limited edition multiples. In the 1960s, Something Else Press also published very famous artists who linked arts, politics, religion, and social structures. Publisher Dick Higgins calls this work the "new mentality," and his press has published pamphlets by Kaprow, Knowles, Corner, Cage, Oldenburg, and George Brecht (*Fig. 9*).

Rubber Stamps, Photography, and Sculpture

The rubber stamp has also been used in publishing artists' books. Claus Beohmler produced *Pinnochio: A Linear Program* in 1969 in Cologne. It shows how to produce an animated Pinnochio using rubber stamps. Two Chilean artists produced a similar book that criticizes Donald Duck with political overtones. It is called *How to Read Donald Duck* (*Fig. 10*).

With the use of photography, many conceptual art pieces have lent themselves to book form. Among the countless examples of artists who have used this form are John Baldessari, who has been making books for many years; Robert Cummings, who illustrates his narratives with hilarious photographs; Mike Mandel, who uses the photographic medium to create baseball cards with artists and photographers, or to explore all the men named Edward Weston in the United States; and Eve Laramee, whose *Fake Art Book* satirizes contemporary artists' publications.

Some artists' books, however, have become works of sculpture, far beyond the norm of what a "book" is. One can play with these books as toys, stand them up as pieces of sculpture, fly them through the air like a paper airplane, excavate for the book in a box of sand, unroll the book like an ancient Chinese scroll, hang the book on the wall like a painting.

10. A forerunner to the rubber stamp movement of the 1970s, Claus Beohmler produced Pinnochio: A Linear Program, *in 1969 in Cologne. Two Chilean artists produced* How to Read Donald Duck.

distribution systems are now established to help disseminate these many small self-published editions.

This artists' book movement of the 1970s was probably predictable in view of the technological advances that permitted the production of inexpensive books and publications that could be distributed directly through the mail and supported by a worldwide system of consumers, archives, critics, exhibitions, catalogs, anthologies, and bibliographies. In fact, with the growth and development of performances by visual artists, sometimes the only documentation for the performance is the data that has been accumulated in book form. Speaking of intermedia, artists throughout the world have used the "flip-book" technique, which makes the book a "movie," or can be considered book as cinema.

The whole process of accessibility on an international scale has led to a network dedicated to the interchange of inexpensive multiples. Many articles have been written about this phenomenon of artists' books, and many exhibitions have taken place since 1971. The accessibility of these works on an international scale has allowed interested persons to come into contact with hundreds of titles, which previously could not have been acquired for any exhibition.

This network works through the mailbox, and the artists' book has truly come into its own as a viable "museum without walls."

Following extensive experience in library science, Judith Hoffberg edited the ARLIS/NA Newsletter from 1972-77. She now publishes and edits Umbrella, *a newsletter on current trends in contemporary art.*

If this seems to be going too far, you must realize that artists today have finally been recognized for their explorations in the book medium. Grants are being offered in some states for artists' publications, bookstores are opening every year in countries throughout the world to distribute these books, and

CAST PAPER

Ann Tullis

1. After straining the water from the pulp to make a thicker consistency, we place the pulp into the mold until the entire relief surface is covered. In this case we used a latex rubber mold with a plaster mothermold backing it. In contrast, the Weege work was done with string attached to a frame and dipped into the pulp.

Editor's introduction: The Institute of Experimental Print-making was founded by Garner Tullis in 1974 in Santa Cruz, California. Over the years, the Institute has opened its facilities to numerous major artists and printers who have produced many of the most excitingly innovative works of paper being made today.

Ann Tullis, who has been the Institute's director since it moved to a 14,000-square-foot warehouse on San Francisco's waterfront, states she is concerned about creating a space for the individual artist. Often, she says, the collaborative process begins with months of conversations between the Institute staff and the artist after which time the varied facilities of the Institute are put to the artist's disposal. The artist can then feel free to play with idea, intuition, and tactile materials.

The Institute is known for its creation of the "unique" rather than the limited edition, as production and systematization are not stressed. Unlike many professional studios, a master printer is not essential to production. Some examples of the varied works produced there are included in Jules Heller's chapter in this book, "Paper as Art: A Contemporary Gallery."

According to Tullis, "At IEP our interest is the artists' interest. We use the medium in any way the artists wish. We are particularly interested in the medium of paper as it is the fastest, most direct, and permanent medium available to the artist. Any artist can put his hands into it, as compared to other print media.

"It is a completely plastic medium, which we have used in varied ways: over strings, as Bill Weege worked; or pressed into molds, as Falkenstein and Nevelson chose to work. Paper pulp can also be modeled directly, though none of our artists thus far has requested this method."

Tullis wrote the captions that accompany the photographs taken at IEP.

2. *The longer the fiber, the better the pulp for these purposes, thus guaranteeing strength. We remove the moisture from the pulp by gently pressing the surface with a sponge.*

3. *After most of the water has been removed and the pulp feels damp, we proceed with a hard pressing with the sponge for surface detail.*

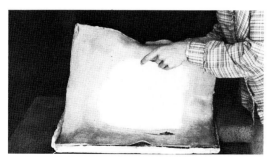

4. *The image is removed after the pulp has dried.*

5. *Finishes vary with each artist. Garner's has been treated with graphite. Nevelson wanted her work sprayed black as if it were wood. Weege watercolored. Bill Dole imbedded gold leaf. Each finish is the artist's prerogative.*

Photos: Bill Netzer

EXPERIENCES WITH PAPER: THE COMMERCIAL WORKSHOP

Kenneth E. Tyler

I can still recall my excitement when, for the first time, I saw very large litho stones in Garo Antreasian's litho department in Indianapolis, Indiana, and the prints Garo had made from those stones. From that day on, I have been involved in large-scale graphics.

After I left my position as technical director of Tamarind, I opened my commercial shop in Los Angeles in 1965 called Gemini Ltd., where I became involved with artists on a quite different level from Tamarind. In January 1966, Gemini G.E.L. was formed, and we began working exclusively as printers and publishers. I think our history has been closely associated with the popular art movement of the 1960s—those artists with whom we collaborated, their particular means of expression at that time, and how the workshop adapted the means to execute their work. It was sort of like building blocks, one project being the foundation for another and so forth until we developed a lot of skill, equipment, and materials to eventually do what we did.

When we look at the growth of our paper program, it began first by convincing a mill to make large paper rolls, then designing and building the presses, then having the artist create the larger prints. Since this system seemed to please the artists we worked with, we kept doing it.

As time went on, it didn't seem very reasonable to me or to the people who collaborated with me that we should just stop with larger paper and presses. We thought we should move into other materials. So we moved into lead embossings with Jasper Johns and into other multiples with Claes Oldenberg, such as castings and movable sculptures. These new experiences gave us more people to work with and gained us new freedom in our expanding research and development programs. Some people thought we were bastardizing the print medium, whereas we felt we were expanding it.

I would like to jot down a few notes regarding my life in printing and papermaking and give some impressions that relate to where we are today at Tyler Graphics Ltd. I remember the early 1960s when handmade paper was available basically only from Europe, and I was aware of no papermakers of my age operating mills in America. Fred Siegenthaler, a papermaker from Basel, sent some handmade papers to Jasper Johns in the mid-1960s. The papers excited Jasper who showed them to me and later I obtained some myself. I believe Tatyana Grosman printed some of Jasper's editions on Siegenthaler's papers. I only used his paper to do experimental proofing. Of the many editions that Gemini published in its early years, handmade paper was used almost exclusively for color trial proofs. These papers rarely became the edition paper.

I don't know whether I can give a clear reasoning for this practice, except that perhaps I did not have a great deal of experience printing on handmade paper and most of the artists seemed to relate to mould-made papers for their editions. Also, I think in many cases the handmade paper of this time was so beautiful and powerful in its surface that it intimidated not only me but also the artists. The images we made at the time looked better on the substrate of a very beautiful Arches or Rives wove paper, and we all found these mould-made papers very dependable for our style of printing. The images we printed needed to be on a neutral ground. Naturally, through the long years of printmaking, many great prints were printed on these French papers, and the history of these prints was well known to us. It took a certain kind of experimentation on our part as well as the artists' to involve ourselves with the exotic handmade papers. (I am leaving all the Japanese papers out of this discussion because they were always around and the workshop used them occasionally.)

The handmade papers that I am speaking about were primarily the waterleaf papers or the unsized papers. Basically all of them were colored papers with pronounced textures. As we got bolder in our printing techniques and in our experimental collaboration with the artists, we began to see the possibilities of making more images on handmade papers. Most of the handmade paper we used during this period (1965–69) was made by Jeff Goodman in Long Island, Siegenthaler from Basel, Laurence Barker and his students from Cranbrook (one of those students, John Koller, eventually became a papermaker on the East Coast with whom we have worked extensively since 1975), Richard de Bas and later on a man in France called Duchene whose papers were introduced in America by Vera Freeman of Andrews/Nelson/Whitehead.

As I examine my use of handmade paper and the involvement with the papermakers I have mentioned, I find all of us are concerned about the state of the handmade craft and where its future will be. In my opinion, handmade and machine-made papers are like brothers and sisters that coexist. I believe they complement each other and I think artists and printmakers would agree with this. I personally would not like to see the printing profession without handmade and very fine mould-made paper, and I hope many more good quality machine-made papers will be introduced in the years to come, in addition to the fine quality Arches and Rives mould-made papers that we presently enjoy using.

In the past 13 years, I have been responsible for publishing (or distilled down further for analysis) approximately 700 editions or 54,000 impressions pulled in various printing techniques of direct flatbed and offset lithography, intaglio, relief, and screen printing, and some unique casting procedures as well as some one-of-a-kind dyed and stained papers. Only 170 of these editions have been on handmade papers, or approximately 12,000 impressions. In other words, less than one-fourth of my production in the last 13 years has been in the area of handmade paper although I have been extensively involved with papermaking since the early 1960s. Presently over 50% of the editions being pulled at Tyler Graphics Ltd. are on handmade paper and the majority of these papers are custom made for the artist and his/her editions.

In part, this change has a great deal to do with the timing of this resurgence and the way it has caught the artists' imagination, much in the same way lithography was affected in the early 1960s and now etching in the 1970s. Since I serve the artist as a collaborator first and secondly as a publisher, I must listen to his or her wishes and try to please them with the widest range of materials possible. If their interest lies in experimentation in handmade paper, obviously, we will accommodate them the best way we can. We will do whatever is necessary to satisfy their needs and make the best project we can.

In 1973 with the assistance of Arjomari and A/N/W, we were able to take Robert Rauschenberg to the Richard de Bas mill in Ambert, France, and make original handmade paper images. (See chapter on the Rauschenberg experiments for details.) Since the Ambert experience, we have continued to explore creative papermaking with other artists. In 1975, Ron Davis created colored papers. That same year, Frank Stella created paper reliefs colored and handpainted at John Koller's mill. Also at Koller's mill, Ellsworth Kelly made a series of colored pulp images in 1976.

It is easier in many ways for a commercial printer, papermaker, or publisher than for a subsidized school facility to spend a considerable amount of money in the area of research and development. Often the money spent in this area is difficult to justify and one's accountant often reminds you of that. It is quite feasible today in any R&D project for experimental papers to cost $100 or more a sheet. I feel the only answer to spending this amount is that if the artist and his collaborators have created something of aesthetic value, then the cost of the materials and activity is not relevant if one can afford to do it. Of course, if out of all of this comes something of no value, as can happen and has at times, it is financially painful.

There are ways to prevent or at least cut down on the expenses of this kind of experimentation for all of us. More technical information can be traded; more professional papermakers and technicians can be asked to advise the craftsman; and certainly the craftsman can use the abundant material available in technical libraries throughout the country. A little bit of technical skill and information with no concern for greater skill and depth in the subject can be both expensive and damaging to the paper movement as we are experiencing it now in this country. I know that without the assistance of many experts we consult, our paper program would not be nearly as successful nor would it have developed as broadly as it has.

There is a wonderful interest by artists in handmade paper now, and those of us involved in making and using it have a responsibility to understand its present role in the creation of art forms and to protect it from becoming a playful and uninteresting craft. After all, none of us here have invented paper-making. We are all here together learning about it and enjoying its rich and varied history and hopefully finding the ways to use it creatively so we will be making a contribution to this majestic and old craft.

I think that there will be increasing emphasis in this country in developing large scale handmade paper, which is the same direction that American printmaking took over a decade ago. I just recently had an experience with John Koller when one of the artists I was working with at his mill felt the need to make larger paper than was possible at Koller's mill. Knowing the limitations of his space and equipment and my keen desire to offer this particular artist the scale he wanted, it was certain that I would need to build my own facility large enough to satisfy the artist's needs. I could not promise Koller sufficient business to warrant his enlarging his equipment and facilities.

Handmade paper in a size 32 by 52 inches and larger is wrought with a great deal of difficulties for the papermaker and is expensive to make. We all know once a large sheet is created for one artist, the next artist may even want it two feet larger and perhaps even calendered or sized, creating even more difficulties.

I wouldn't like to see a rapid increase in paper size, as it happened in lithography, only to find the interest in extra large scale is limited. It is interesting to speculate that perhaps the papermaker of today serving the art community can learn from the experiences lithographers have had regarding oversize prints and equipment and the many physical problems, bad backs, etc. that developed from working on large machines and prints. I find this speculation quite fascinating.

Kenneth E. Tyler is the owner of Tyler Graphics Ltd., in Bedford Village, New York.

TWO RAUSCHENBERG PAPER PROJECTS

Kenneth E. Tyler and Rosamund Felsen

Editor's Note: In March of 1974, Robert Rauschenberg went to the Richard de Bas paper mill in France to create a series of paper pieces. He was accompanied by Kenneth Tyler, then of Gemini G.E.L.

Although artists like Alan Shields, William Weege, Joel Fisher, and Garner Tullis had been working in the paper medium since the late 1960s, Rauschenberg's example did much to bring about the widespread attention to the art and craft of papermaking that we see today, largely due to his stature and leadership in the art world, as well as the innovative processes he explores. Rauschenberg, dedicated to breaking with tradition, used two- and three-dimensional processes to produce multiples that combine printmaking and sculpture.

Again in 1975, Rauschenberg and a group from Gemini went to Ahmedabad, India, to work on a project inspired by the unique paper and fabric processes used there.

Tyler wrote the following report on the visit to France, which is followed by Rosamund Felsen's report on the visit to India. Although these previously unpublished reports were not included in the 1978 paper conference in San Francisco, we at the World Print Council thought the information would be of great interest to paper lovers.

Rauschenberg's Trip to France

One of the very first usages of handmade paper that gained a great deal of attention in the art world was the Rauschenberg project in Ambert, France.

Planning for it began soon after my visit to Japan in 1972. I reflected on the paper works I had seen in Kyoto and began considering the possibilities of working directly with artists in a hand papermill. Not knowing what direction this collaboration would take, I felt that the project should be involved in color papermaking in a European mill. I broached the idea with Vera Freeman of A/N/W and Elie d'Humieres of Arjomari who enthusiastically pledged their support and assistance. By early 1973, Elie d'Humieres had arranged for me to procure the Richard de Bas mill with three papermakers for one week in August of that year.

I contacted Robert Rauschenberg and explained the proposed project in detail. He became very excited with the possibilities and quickly began working on ideas that would use colored pulp with objects and suggested I experiment with incorporating printed tissue paper onto or in handmade papers. We developed four-color process screens from collage

Photo: Gianfranco Gorgoni

1. Tyler pours yellow pulp in tin mould for Bit. *Rauschenberg observes. Richard de Bas mill, Ambert, France. August, 1973. Photos 1 through 4 courtesy Tyler Graphics Ltd., Bedford Village, New York.*

2. Papermaker at Richard de Bas mill prepares to couch Bit *paper images. Kay Tyler cuts printed tissue papers for the project. 1973.*

Photo: Gianfranco Gorgoni

3. End view of the 1300s screw press at the Richard de Bas mill.

material the artist supplied, and we printed the images with water base inks onto special machine-made tissue paper that Bernard Guerlain of Special Papers obtained for us.

The paper and ink we selected permitted the embedding of the printed tissue papers into the newly made papers and caused the water soluble ink to bleed slightly. After pressing, the thin tissue papers became an integral part of the base paper. These printed papers were made in our workshop and shipped to Ambert, France, prior to our arrival.

We informed the director of the mill, Marius Peraudeau, of our intention to create shaped colored papers embedding printed tissue paper and other foreign materials. He equipped his mill with special Swiss water-soluble dies to color the cotton pulp and arranged to have a local tinsmith available to construct shaped metal moulds from the artist's designs.

During the four days at the mill, Rauschenberg created 12 different images for editioning. Five of these images were formed without the use of tin moulds, colored pulp, or laminated printed tissue paper and consisted of off-white duplex sheets with embedded cord, twine, and old remnants of printed cloth. These five editions were titled *Pages 1 to 5.*

The seven images titled *Roan, Hind, Scow, Bit, Vale, Vale State I,* and *Link* were single sheets formed on wove paper moulds with tin image moulds placed on top, serving as compartments to receive the poured liquid color pulp. Each embedded printed tissue paper was placed by hand onto the wet-colored pulp before the paper was pressed between felts in the mill's early 14th century wooden screw press.

Two of the mill's papermakers made the wove mould paper for the duplex sheets and couched all of the color pulp works. The artist directed me in the mixing of all of his colored pulp, and I assisted him in embedding the printed tissue papers. With the aid of Vera Freeman and Elie d'Humieres as translators, I directed and supervised the total project assisted by Kay Tyler and Bob Peterson and the two French papermakers. Kay Tyler kept records of all procedures so the mill could be instructed in producing the balance of our papers after our depature.

Approximately one-third of the 300 papers made were created during these four days. It was the intention of both the artist and myself that each paper made would relate to a stan-

4. Tyler positions tin mould for one of the **Pages and Fuses** *works as Rauschenberg prepares location for the printed tissue papers. Richard de Bas Mill. 1973.*

5. (Right) Robert Rauschenberg. Vale. 1973. Handmade paper, silkscreen on Japanese tissue. 19½" × 24¾". Courtesy Gemini G.E.L., Los Angeles. Vale is natural-fiber handmade paper, pigmented, and formed in a mould. Sheets of Japanese tissue paper were screen printed with magazine images and placed on the newly formed wet sheets of paper, pressed and left to dry naturally.

dard work, which served as the model for each edition, and still permit each paper made to have variation. This became important during the project since many papers were to be made from the standard after we left France.

This project represented the first time an artist collaborated with me in a papermill and signaled, I think, the resurgence of hand papermaking as an art form in this country.

6. *Carpenters lay bamboo strips on* Pit Boss. *Ahmedabad, India. 1975. Photos 5 through 10 courtesy Gemini G.E.L., Los Angeles.*

7. *Woman prepares rag-mud for* Unions. *India. 1975.*

8. *Kilji (left) and Ram (right) pour paper pulp for* Bones and Unions. *India. 1975.*

9. *Rauschenberg places fabric squares on* Hard Eight. *India. 1975.*

Robert Rauschenberg had been invited by a wealthy Indian family to come to Ahmedabad, India, to work on an art project of his own design.

This particular family had a long time interest in contemporary American art and had come to know such luminaries as John Cage, Jasper Johns, and Rauschenberg. Some years before, they had invited the Merce Cunningham Dance Troupe to Ahmedabad for a stop on their world tour.

Rauschenberg thought something particularly interesting might develop if this invitation expanded to become a Gemini project. Since his initial interest in papermaking had been whetted by the earlier Gemini project at the Richard de Bas mill in Ambert, France, the invitation was accepted. Arrangements were made to have a facility available for a papermaking project in Ahmedabad. It turned out to be a facility of historical importance.

Ahmedabad was the birthplace of Mahatma Gandhi. It was here in 1917 that Gandhi's first nonviolent demonstration occurred. Ahmedabad was, and still is, a large textile center in northwest India, about 400 miles north of Bombay. The 1917 strike Gandhi organized was against the owners of a textile factory on behalf of the workers. It was in Ahmedabad that Gandhi developed his passive nonviolent philosophy in order to uplift the condition of the people who had long been called the "untouchables."

Among the overwhelming problems concerning this poorest "caste" of the Indian people was the fact that they were totally unskilled in any work except the lowliest. Historically, they had not been allowed to learn anything that would enable them to upgrade their position in the caste system. Gandhi's plan was to establish craft compounds. He set up ashrams where these people could learn to work with leather, cabinetry, papermaking, and other crafts. The facility made available to us for the Rauschenberg/Gemini project was one of these ashrams.

We arrived for the project in May 1975, a particularly significant moment in Indian political history. Events were shaping toward the downfall of Indira Gandhi, who was just beginning to overstep her bounds. As guests of this wealthy

10. Robert Rauschenberg. Capitol, *from* Bones and Unions. *1975.*
Rag-mud, bamboo, silk, string, glass, teak wood. 34″ × 53½″ × 4″.

amily, we were accommodated in a Le Corbusier house, complete with swimming pool (an essential, as our stay in Ahmedabad immediately preceded the annual monsoon season and this
was probably the most difficult season for us to be there as the
temperature every day exceeded 100 degrees with the humidity
coming close to 100%; we left as soon as the monsoon came).

The project lasted a month. With Rauschenberg were his
son Christopher, Robert Petersen, and Hisashika Takahashi.
My husband Sid and I were accompanied by our daughter
Suzy, at that time 13, and Charly Ritt, long-time Gemini printer
and collaborator, and Gianfranco Gorgoni, photographer.

As this was the textile capital of India, the first thing to
do was to go to the textile mill and select fabrics to be used in
the project. A large selection of cotton prints was chosen by
Rauschenberg. A guest house had been set up as a studio. It
was here that Rauschenberg made the drawings for the five
paper pieces. The plan was to inlay into the paper pieces strips
of bamboo, which were available in great abundance. Pre-cut

pieces of fabric would then be placed between the tied bamboo
strips.

We then went to the Ashram and met the people who
worked on the papermill. The two papermakers we worked
with were Ram and Khilji. Women worked in the mill, too,
but their responsibilities were confined to sorting papers and
carrying bundles of rags upstairs to the loft where the large
pulpmaker was located. When ready, the pulp was brought
downstairs and kept in cement vats. Water overflowed onto the
floor around the working area.

In the actual process of making the Rauschenberg pieces,
a single layer of pulp was poured into the screen. The tied
bamboo strips were then laid on top of the pulp, and then the
fabric pieces were put into place. A second layer of pulp followed. Each piece was placed between Army blankets (felts)
and couched. The pieces were then set out to dry on racks.

Their presses were adorned with what we know as swasti-
kas, but to the Indians it is a good luck sign. We completed the

editions of *Bones* while Rauschenberg was already developing the next part of the project.

Back at the studio, Rauschenberg had begun experimentation on a substance inspired by the traditional adobe-like material the Indians have used for building their houses. Somehow, he wanted to incorporate the use of the paper pulp. With the advice of a craftsman named Pal Balbu, he created a substance that combined the "adobe mud" and paper pulp.

The tradition in using this material was to add certain ingredients to the adobe as an insect repellant. These aromatic spices, fenugreek powder, and ground tamarind seed were mixed with chalk powder, gum powder, and copper sulphate. Gum powder and tamarind seed were mixed to a paste with water and then added to the mud. Rauschenberg designated this substance "rag-mud."

This material was to be used to make the sculptural pieces he had been yearning to make as part of the paper project. While it is true the *Bones* were somewhat relief-like, the next half of the project, dubbed *Unions,* was truly three-dimensional.

Fashioning the "rag-mud" into shapes by hand, Rauschenberg incorporated the use of rope, twine, bamboo, and brilliantly colored silk. Prototypes were developed for six editions.

We worked outside on a large flat table or on the slate patio floor. Because of the high humidity, precautions were taken to prevent mildew and insect infestation. It wasn't possible to allow the pieces to dry naturally as the mildew would develop overnight and the insects would arrive shortly thereafter. To dry the pieces as quickly as possible, they were driven in a converted ambulance to the textile factory and placed in ovens (used for drying screened textiles) for several hours until the one-and-a-half-inch to three-inch thick rag-mud was thoroughly dried.

Certainly one of the most intriguing aspects of this project was working in an environment with far fewer technical re-sources at our disposal. Rauschenberg resorted to the use of the most basic information in a cultural situation so very different from our own. The members of the family with whom we were staying had been educated in England and were highly sophisticated. The workers in the paper mill and the carpenters and craftsmen who worked with us on the project were helpful, hardworking, and responsible, even though there was no common language between us.

Rauschenberg suggested that we give a party one evening for the families of all who had worked on the project. It is typical of him in his all-encompassing desire to extend art universally. All the completed pieces from *Bones* and *Unions* were on display. Also invited were the friends of our hosts. This combination of guests was a unique experience, we were told. Rauschenberg showed a film about his work which included his well-known performance/dance pieces. Everyone sat on Indian blankets spread on the lawn and watched intently.

Thirty days had elapsed. The *Unions* editions were to be completed by Charly Ritt who was to stay on after the rest of us left for other commitments. The *Unions* were completed under his supervision by early July.

Kenneth E. Tyler served as technical director of the Tamarind Lithography Workshop from 1964-65. In 1965 he founded Gemini Ltd., in Los Angeles. He served as director and master printer there and at Gemini G.E.L. until 1974 when he founded Tyler Graphics Ltd., in Bedford Village, New York, where he continues to collaborate with such artists as Stella, Kelly, Hockney, and many others.

Rosamund Felsen, part owner of Gemini G.E.L., in Los Angeles, has worked as curator of prints at the Pasadena Art Museum and is currently owner of the Rosamund Felsen Gallery in Los Angeles.

EXPERIMENTAL APPROACHES TO PAPER IN ART

A Panel Discussion

Riva Castleman
Robert Flynn Johnson
Kenneth Noland
Garner Tullis
Kenneth E. Tyler

Castleman: As a word of introduction, I want to say that artists just didn't begin to experiment with paper. After all, whenever anyone makes the choice of paper—whether it's the bookmaker, the artist, or the printer—the choice is obviously of an experimental nature until you see how it comes out. Rather than go all all the way back and discuss how artists in previous centuries have experimented with paper, we'll sort of plunge into the 20th century.

One of the things that has continued into the 20th century is the use of several kinds of paper for one image in printmaking. This sort of tradition was not particularly experimental. It was a tradition based on economics. So we have to be a little wary about identifying the paper behind a print as being exactly what was the ideal background for that work. For instance, the limited editions of books from France, illustrated by such people as Picasso, Miró, and Matisse, are very often on three or four different kinds of paper, some on very fine Japan imperial and other fine papers, which was not necessarily the way the artist saw his or her work.

Experimental papers for printed art certainly have found experimenters among others than the artist. I want to point out that Picasso prints of the 1930s were very often printed on vellum, which is a very unnatural medium for etching as the ink tends to pop off, and the material, which is so neutral, seems to move around. This isn't paper however, so we don't have to worry about it.

Another example is of a very interesting and experimental publisher named Ilya Zdanevich of Paris, who was among the first to locate and work with the papers of Richard de Bas. I think it is because of his work that Americans know about this handmade paper.

Before I open this up to the panel, I'd like to read the following quotation I especially like because I'm interested in prints and perhaps less in the sculptural areas of paper. Steven Kasher wrote the passage about paper in a recent article in *Artforum*. He took a very strong stance when he penned this remark: "Paper is primarily a surface for recording information. As such it exists to be obscured. It aspires to whiteness only in order to make its own obscuring more poignant. And the text it supports is what presses paper completely into the two-dimensional. Only an empty sheet of paper has volume."

Without further ado, I'd like Robert Johnson to relate a little of the history of experimentation in paper, in printing particularly.

Johnson: The fact that Riva Castleman and I are both curators of museum collections separates us in a certain sense from the other three panelists, who are involved in the creation, publication, and printing of works of art. As curators, we are very much involved in a future step, which is the interpretation, exhibition, and acquisition of works of art for museum collections. From a curatorial point of view, the reawakening of interest in varied forms of paper as an integral support for a drawn or printed image, or created in such a manner to become artwork in itself is an exciting and challenging event, which has occurred at a quickening pace in the past 15 years.

But, to reflect for a moment, I'd like to discuss the early part of the 20th century in terms of the use of paper. Except for a few beautiful Max Weber color wooducts of *circa* 1918 on Japanese paper, also used by such artists as Louis Shankar and Ann Ryan in the 1930s and 1940s for color woodcuts, most American prints before 1950 were small and black-and-white.

The 1950s brought along a gradual revival of interest in prints after the lack of interest caused during the Depression and the poor economic conditions that followed. This reawakening of interest in prints was caused by: the strong influence of Stanley William Hayter and his intaglio techniques; artists such as Leonard Baskin with his extraordinary, over-life-sized woodcuts and his interest in wood engraving; and finally and possibly most important, the resurgence of lithography through the efforts of many in the late 1950s.

However, except for few experimental paperworks by Jackson Pollock on paper made by Douglass Howell and occasional prints, usually woodcuts or wood engravings on Japanese paper, it is only in the late 1950s and early 1960s that editions of prints on beautiful handmade papers began appearing, published by Tatyana Grosman at the Universal Limited Art Editions on Long Island.

Artists sought out and acquired specific papers for specific prints. Rather than merely a printing surface, the paper was becoming an integral part of the art itself. Thus, a Frankenthaler color lithograph would be printed on a green-brown handmade paper. Jim Dine printed his memorable lithograph, *White Teeth,* on black Fabriano; and Larry Rivers produced the cover of Universal Limited's first book, *Stones,* as a folded thick piece of handmade paper made out of old blue jeans. The aesthetics of paper and of the print were becoming brilliantly intertwined at ULAE.

Because of the scarcity of handmade paper then and the lack of technology to make large-scale sheets of paper, however, many artists, notably Robert Rauschenberg, continued to have to use conventional types of paper in their printmaking. Significantly, in the 1970s, the notion has been gradually slipping away that the ideal of a good edition of 100 prints is that they must necessarily all rigidly be the same and, if any of the impressions do not meet up to the *bon à tirer* impression, they should be torn up as unsuitable. It is often just those so-called imperfect proofs, possibly on odd-sized paper or color paper, that can be the most intriguing to the artist and, I might add, also to the curator. The manipulated print is one that has the greatest so-called touch of the artist within it and brings the print collector closest to the creativity that made it.

Too many of the prints created in the 1960s, especially lithographs and serigraphs, were slickly and coldly produced to fulfill the art-buying need of a public hungry for the thrill of owning modern art. Many of the prints produced at the time were created with little or no thought of the paper as anything more than a nonobtrusive surface to carry a recognizable image, preferably in color. To be eventually purchased by the public, they were considered not so much as prints but as surrogate paintings to fulfill the needs of an art public that in large part could not afford to buy the paintings and drawings of those artists.

Happily this trend is changing, as many in the art public have become more knowledgeable about art and are demanding much more in their choice of works of art on paper. Two of the major art magazines continue to pollute with color-print ads for Grandma Moses-style prints. But in ever increasing numbers, progressive workshops and publishers are expanding and exploring a wide area of artistic expression using whatever materials are necessary for the completion of the artist's idea. We happily live in an artistic era that more often asks the question "why not" than "why." Because of this, works such as those of Frankenthaler, Stella, Noland, and others are part of public and private art collections today.

The Idea of "Experimentation"

Castleman: Will Garner Tullis please tell the audience a little of his opinion of the current scene?

Tullis: I'd like to address the idea of the word "experimentation" and the idea of making art. As Whistler says, "When art happens, no hovel is safe from it, no king can depend on it." Art happens out of limits. While form is born from limits, experimentation is having the astuteness to define one's own limits.

As I see it, one's limits come from hard intellectual posturing that makes hard ideas and art. If the vehicle (medium) can be part and parcel of the image and metaphysically harmonious with it, then experimentation with the medium can be justified. If you've made the image at the same time that you've made the vehicle, then that totality can be an exciting piece of art. It was always these concerns that brought me to experiment and to try to excite others to experiment.

Eugene Feldman first introduced the idea "experimental" to me. He called himself an experimental printmaker. He wanted to make offset prints though offset was a dirty word in the print business in the late 1950s and early 1960s. To study with him, to work in, and literally to sweep his shop, as I did at first, I had to agree that I would take up the cause of experimental printmaking. It was out of love for him that I named my shop the Institute of Experimental Printmaking.

Experimental is a very dangerous word. When one normally thinks of experiment, in the popular sense of the word, one thinks of patents for inventions and of who did it when. I've finally grown up to realize it really doesn't much matter. What's important is what really happens. Our egos tend to get in the way, and they really warp our sense of defining our own limits if we're trying to get on the ground floor of a new idea.

The use of dimensionally formed paper began as early as the 15th century with the blind embossments that were proofing of the stamping of armor in the city of Florence. In the 18th century, a Japanese printmaker, Suzuki Horunobu, was interested in very subtle embossment and produced the gauffrage print.

In the late 19th and early 20th centuries, three French sculptors—Aristide Rousseau, Maurice Dumond, and Pierre Roche—embossed prints from plaster and steel molds. Why did they do that? Why were they trying to force what was normally a flat sheet of paper into a dimensional form? Because their needs for themselves and their identities were formed from the viewpoint of sculpture, and things in the round and *bas relief* as opposed to flat. So they sought means to give a paper image dimensionality. It wasn't that paper was something new and exciting. It was just that it served a specific need, an intellectual demand that would enhance and thus state what they wanted to state in their work, which is what the art is about.

You can walk away from a work of art and still carry that art with you. What you carry is the idea of that work of art, not who invented the process, where he found the material, or really what the material is. It's the idea that's most important from my point of view.

With regard to experimentation, I think the German philosopher Goethe was quite right when he said that one should find fault with any person of the past only on your knees. I think what he's really talking about is that we climb over the backs of all those who have gone before us and try to see a little further with our eyes and our minds. But it's all built on the base of all that work of discovery, all that mental work before.

What we do when we experiment is to do new things in an old way and old things in a new way.

Publishers and Experimentation

Tyler: The one positive difference I can perhaps add to this discussion is that, as publishers of prints, we have always considered paper to be one of our most primary materials. We were always satisfying the artist and his search or quest for making imagery in our printing atelier.

As time went on, we went from lithography to screen, from screen to intaglio, and from woodcut and into a lot of mixed media arrangements. We've also gone from the traditional flatbed, direct hand printing press to even the direct flatbed that is used in Europe, called the mailander press.

We have often felt that if the workshop environment for an

artist were kept open and experimental, the art work that would come out would be exciting and would always increase the dimension of the artist in his image making. So, as collaborators, we always first considered that the paper was basically the substrate for the printed image and felt we could take it into the third dimension by embossing it and moving it around.

As we began using more and more handmade paper in the last several years, we saw it as plain simple molded image that could be painted on or attached with collage materials that may or may not have been previously printed on. We felt that the paper itself had the possibility of becoming a painting medium. Previously, print publishers had been expected to make very uniform and even editions and were not supposed to get into variants or unique individual pieces.

Of course, using paper as a medium brings up a whole lot of new questions in the area of editioning. We question whether paper works should be signed and numbered as prints, or whether they might have letters, alphabets or just family names. I'd rather leave that up to everyone to solve the issue for themselves.

We've made our decision, at Tyler Graphics Ltd., that we are trying not to call them editions, and we may or may not sign and number them in the conventional way.

There is a myth about handmade paper that I'd like to challenge. The myth is that handmade paper is necessarily dimensionally unstable and necessarily forces the printer to print uneven editions. I would like to suggest to you very strongly that the prints at our workshop on handmade paper are even and uniform, and we do whatever is necessary to accomplish that.

Since we print in all media, from lithography to screen, we are aware that some papers work better in some areas than in others. But I assure you that if you have the tenacity to stick with handmade paper, you can learn a great deal about it. At first, I think, people become discouraged because it is so different from mould-made in its ink absorption and its surface.

The other factor I'd like to raise is economics. Traditionally, we have worked with mould-made paper manufacturers by developing, assisting and suggesting the areas in which their papers could possibly be improved for our industry or craft. Some of it has come true because there actually has been

enough tonnage there, and it was economically feasible for the mills to partake in a new development.

As time went on, artists would seek from us new colored papers that were not economically feasible in small lots from the mould-made paper industry. Each mill has a minimum tonnage for special papermaking. If an artist only wanted to do one or two editions, that would amount to no more than several hundred pieces of art in total. Obviously, we could not invest in having tons of paper on hand that perhaps would never be used again.

This is a very important point to mention because we found with the handmade paper mills, we were able to get small batches of various unique papers whenever we wanted them. This, of course, made it possible for us to do a lot of things with the artist in the area of color and weight and surface that hitherto was not possibile. The distance between Europe and us has always been a financial one as well as one of time. As you can appreciate, to ship materials a long distance is very expensive.

Finally, I want to reiterate that I think it is important to have access to both types of paper. Machine and mould-made papers can and should exist side by side with handmade papers.

The Importance of Handling Pulp

Noland: I didn't get interested in paper until, in effect, I got my hands wet making some a couple of years ago at the Institute of Experimental Printmaking. I found, in handling the stuff of paper, the pulp, that it was something I could handle in the same way as wet paint and raw cotton duck.

In modern art, there's been a gradual elimination of using drawing to make images. As that has been withdrawn by some artists, a different kind of emphasis has taken place in drawing. It has more to do with tactile relations in handling stuff. I am referring to the interest in collages as an example—taking things and pasting and sticking them instead of having to depict things by drawing.

This new interest in papermaking gives artists access to the stuff. There is something that hovers around the threshold of surface, and this new interest in papermaking doesn't make you have to accept the surface. You can get underneath it rather than assume that you have to accept the surface and put something on it. You can get in it and make something come up and become the surface.

That's what I'm interested in exploring. It is a nice relationship between the water and the fibers and that kind of elusive glueyness as you handle it. It is a very delicate area and especially when you take rags that have been bleached by the sun and washed a lot. To have access to that in the making of paper is very exciting to me.

Recycling

Castleman: Ken Noland's reference to collage is extremely important as I think the 20th century is obviously a period of recycling—using materials that exist in other forms over again. A good part of this conference was devoted to preserving what we have, but recycling is obviously the other side of the coin. Paper itself is the great recycling medium as it exists in its classical form. I wonder whether just the idea of recycling, not necessarily remaking old materials or old prints, but as it evolves from collage, the reuse of some material can become the basis for experimentation.

Tyler: I could see where somebody could make something, start taking it apart and then reposition it into a new work. Some artists have, in fact, taken already completed works of art, recycled them, and made pieces of paper on which they then made new pieces of art. They are now numbering and editioning the works by the number of times they have been recycled!

I can't predict what is going to happen. There might be a lifetime of recycling the same piece of paper and reworking it.

A project we worked with at Tyler Graphics Ltd., the Frank Stella project, utilized the concept of reuse in the way colored pulps were mixed together to form new colors. The darkish gray-black pulp used in the later pieces is really the end of the project since *all* the colors that were used were put into one large vat to make that particular color. We would never be able to make that color again because we don't know what we put in there.

Tullis: Recycling became important to me intellectually because I wanted to take an original painting, and I wanted to make an original print to defy the law that said I couldn't, the law from the Print Council of America. I did numbers of them; I numbered them 1/20th of one, 2/20ths of one, 3/20ths of one, because they came from one painting and all of the painting was in every piece of paper. That wasn't really an effort to recycle a painting. It was just a reaction against the idea that someone else could tell an artist what they could or couldn't do with a print.

I think recycling becomes important in a metaphysical or spiritual way, too. It's like having the whole journey in the work of art when it's done. In the painting process, we have to view the work through X ray to understand its evolution. If we want to know how a sculpture was made, we have to have been there when it was made. If we see a video of the sculpture process, all we really see is the video since the video isn't really the sculpture process.

Sam Francis addresses the idea of the development of an idea through time, in a different way. The series of prints is called *A Fixed Course of Changes.* He started with a set of limits and a set of physical problems, such as viscosity, mixing oil and water and making a monotype with certain colors. The idea develops through the changing of the plates. It's a series, and you can see how Francis started with his first print and ended with the last, which is the final resolution of that idea. In that series of prints, which should be exhibited together, one is able to see how an artist's mind functions through original works of art in their evolving state. That's a type of recycling of thought.

I'm disinterested in doing anything that's editioning in the classical way. For me, that's commercial. If I'm involved in experimentation as an artist and I'm going to help other artists to experiment, I want to resolve an idea and go on to the next idea. I don't want to stop. I'm jealous of time in that way.

Castleman: When one works in a medium such as prints or sculpture that requires a mold, it seems the choice of making an edition is inherently demanded. What would you, Robert Johnson, think of a work that could be editioned? Is it more acceptable it if is in an edition? And would you consider all these works to be in the print area because they are paper and cast?

Johnson: The whole question of the edition is one that I am personally most pleased that we are starting to get away from, especially from the multiple print that is run out in 100 or 200 impressions, each one which is pristine and all that.

I prefer the idea of the artist's touch. As an art historian-curator with a collection at hand, I can go back and see how Whistler lovingly printed his beautiful Venetian etchings, carefully hand wiping various impressions with no thought of editioning them or having each one the same. Whistler, as well as others, simply rolled up his sleeves and made works of art. He was not involved very much in the commercial aspects of it.

There are problems with this approach, albeit a commercial fact of life. For example, someone may go to the Boston Museum of Fine Arts or the New York Museum of Modern Art and see a particular print, let's say a two-color etching with some hand wiping on it. That person may then go to his or her local art gallery and say, "I want an impression of that edition of 50 just like the one in the Boston Museum of Fine Arts." What invariably happens is they pull one from the drawer, and it doesn't have as much wiping, or it has more. The collector is aghast and says, "Well, I want one just like theirs because if I don't have one just like theirs, more than likely the museum has made a selection that was the best and this one is not as good."

Frequently, we get into this whole level of qualitative relationships between prints that one doesn't have in drawings. Each drawing is different. One doesn't have this problem in painting. But for some reason, when it comes into the print-making area or the printmaking multiple area with handmade papers, suddenly, there is this idea that it has to be just like the next one.

Editioning

As a curator, I like the idea of the touch involved in the work of art, the sense of the artist and the printer working in collaboration to produce an image. I have, however, one suggestion that I think might be helpful for artists and publishers who are making prints that vary slightly in terms of color or in terms of size if the margins are slightly different. It is that the edition not be numbered 1/30, 2/30, 3/30, giving the uninitiated the sense that each of the 30 is the same. It might be better to call them, "version 1/30," "version 2/30," to differentiate the fact that there are 30 of those prints or multiples but that each one varies to a certain extent. I think this might possibly clear up the situation.

But I am very intrigued as a curator, and I think that the curators of museums have to take the art as it comes and should not set up ground rules like those of stamp collecting that puts our artists on the spot. You have to give them their freedom.

Castleman: Ken Noland, have you ever faced that particular problem? Have you ever thought that as you get involved in paper, you might be getting involved in the making of multiple art? What has this meant to you?

Noland: I think that the fact that we're all here means there is a new access to touch. Everybody's very excited about getting at the making of paper. David Smith, the sculptor, was a good friend of mine. He wouldn't make multiples. He wanted to put in *each* thing he made the possibility of a success or failure. He felt that that was what making art was about—that you've got to take your chances.

I think that artists in the 1960s who stuck to making their own sculpture, painting, and other art forms wanted to keep that possibility of success and failure constantly present. Not that it wouldn't necessarily happen in making multiples, but the concept of editioning does eliminate some of the choices and decisions that could be made. It generalizes a judgment that you may have to make when you are making unique things.

I believe that making art is a practice. It's a lifetime practice. Once you start making art, you have to keep using your hands, and you use your hands all your life. You gain skills, and you also have to set up things that are out of the control of your hands to challenge your skills.

I think things *can* be repeated. When it comes to a choice of things that I prefer, I'd rather have a thing that has that characteristic of *repeatability*, though I can't understand making 100 prints that are mechanically uniform. Making just plain straight paper would be that way, too, wouldn't it? For example, the way the Japanese can cast sheet after sheet of beautiful paper. It's such a pleasure to take a piece of Japanese paper and hold it to the light and see that beautiful, elegant degree of control. Of course, that doesn't make art; it makes a piece of paper. It doesn't make art because there are some other kinds of decisions and choices where the possibility of failure and success enters it. That possibility is what interests me about making art.

Question from Audience: How far are you, as caretakers of art, willing to go if, even after I've finished the work, I as an artist am interested in a growth and change process, even to the point of deterioration?

Paper from Artichokes

Tullis: A friend of mine did that. He made paper from artichokes. We had some printers from Gemini come to our shop (at the Institute for Experimental Printmaking) to learn how to make paper. One of the people who was helping me with the workshop, Charlie Horton, found one of the printers from Gemini making paper out of fresh artichokes. He realized there was a bug that ate artichokes, so he made a conceptual print that was several of these bugs with a sheet of artichoke paper in a clear plexiglass box. The bugs began to eat, and the pattern was quite beautiful, but eventually the paper was totally consumed. Would you be interested in acquiring this work of art?

Johnson: It depends on which point in time. Curators, especially when we are talking to groups of art students, plead a lot. Sometimes we beg or bring in our paper conservators to also give a few choruses of "Please use good materials."

On the other hand, there are two ways of using bad materials. One is created by necessity because the artist can't afford good enough materials. The second, the one the question was pointed at, is when the artist has the option of all materials and actually chooses materials that are by nature ephemeral, or the artist actually plans parts within the print or structure that are destructive to other parts. This is, again, where I think that our role as curators is not to tell artists what to do but to receive the works of art and deal with them at that point. The only thing that one can do is try to preserve the work of art, and this might sound strange, but to try to preserve the work of art very much at the point at which one acquires it.

For instance, Franz Kline did a number of very beautiful black ink wash drawings in the 1950s on a kind of ivory newsprint in use at the time. You can very well imagine what newsprint looks like 25 years later if it hasn't been deacidified or worked on. The Kline drawings have now turned black ink

on a kind of orange paper. Whether that was his intention is a problem curators have to work with. Our conception, of course, is to try to stabilize it as much as possible.

But if artists are determined to use degradable-type materials, one just has to deal as best one can. There is no law against artists such as Ruscha using raspberry jam or Pepto-Bismol in their works of art. It is not so much the curator's problem, but when we take it back to the laboratory and put it on the desk of our paper conservator, that's when the problems really begin.

Castleman: I'd like to answer the question further because the point behind the question wasn't the destruction of the work itself. It was our attitude about open editions that seems to have been closed. For myself, I don't care, because, in truth, I'm interested in the art. I'm not interested in the economics of whether the artist is making 30, 100, or one. If only one is the good one, I want that one.

Sure, it's hard on the historian, or the person who keeps the data in the computer, as to how to describe all this. But language is a marvelous thing. Soon we're substituting numbers and all sorts of other abstract notions for our language.

So, I think our language grows because the artist confronts us with a whole lot of new things. As far as curators are concerned, I'm sure we can easily state we have no interest whatsoever in upholding what is literally an economic problem.

Uniform and Nonuniform Editions

Question from Audience: I feel that the panel is missing the sensitivity that is part of editioning a uniform edition of prints, that in the hot defense of nonuniform editions, not enough credit is being given to the more subtle qualities of the uniform edition. The amount of uniformity is, after all, relative.

Tyler: There is a lot of sensitivity in printing uniform edition work, and I don't think anyone in this audience wants to suggest we suddenly spin everything down and stop doing that. I think it's going to coexist forever.

Obviously, people who made monotypes many years ago were also making uniform editions. In the thinking process, the artist is forever changing. We are talking about imagery; we're talking about *art;* we're not talking about restriction.

We all remember the good old days when we had rule

books that told us that mixed media, monotype, et cetera, et cetera, wasn't allowed in that show. I hope that's ended. It's comforting to know that museum people are out there supporting all the new adventures in almost every kind of way that happens today.

It's a very difficult subject, which I think you get nowhere with because I believe that the state of the art in all printing today is what it is because there have been an awful lot of people that have labored and toiled to learn their craft. They have become very sophisticated, and they have a purpose and a dedication. We should never knock that. We should support it.

Tullis: I collect prints as well as make prints. I collect old prints, and they are all editioned prints. I think I fell in love with printmaking from editioned prints of uniformity, not from individual prints.

What I'm expressing about the unique has to do with me as a studio artist addressing need at my time and perhaps facilitating other artists' needs at their time to grow rapidly intellectually. But I would also stand with Ken Tyler and fight for anyone who is at a point to do uniform prints.

Prints have a proud history. The first limited prints, if I remember, begin with Vollard. He did a portrait of his dealer when he did it. I think of that Renoir that I bought because I love it and because it stands for that idea that the first limited print edition with uniformity was 1,000, not 20 or 27.

Dürer traveled the countryside. He didn't number his prints. When he saw somebody he liked or he needed food, he printed a print. Prints were for people; prints were accessible; prints were populace-oriented.

Johnson: Here in San Francisco at the collection of the Achenbach Foundation for Graphic Arts, which has only been here for less than 30 years, we have examples of works by Mantegna, Dürer, and Schongauer that were printed in multiple editions. If the artists had all worked in a very experimental singular way, we would not have any of these works. They would all be in the Albertina or in the British Museum. We do have monotypes, for instance, by Pissarro, Degas, and others.

One of the things that is so gratifying, so optimistic about the 1970s is not that somehow that edition prints are on the outs, or unique prints or manipulated prints are in, but that there is an opening up. There are different ways of going about the making of art that will be acceptable—the monotype, the collage print, and so on. That is the thing that's so gratifying.

In terms of edition prints, one of the great potentials about technology is that just recently one could buy an original Jasper Johns serigraph in an edition of 3,000 for only $10. It's probably the only Jasper Johns that a great number of people will be able to own in this day and age, but here one can own it. It's only because we have the technology to make a beautiful, unlimited, so to speak, original print for anybody that has the very small amount of money needed to acquire it.

Castleman: Certainly at the Museum of Modern Art in New York, we are seeing great quantities of work both multiple and unique works of paper. We have had a difficulty in categorizing conceptual art and various other movements in the last decade or so. We've also had difficulty in placing cast paper, and multiple paper. In fact, I would say that every day there is something new, which just shows that it has been a long time since paper has been dealt with creatively. It's fascinating to see.

Collaboration

Question from Audience: Please comment on the concept of collaboration.

Tullis: My attitude has always been that when you ask somebody else for help, a collaboration has begun. The whole point of helping anybody is to enable them. When you set up a shop or a mill or invite someone into your studio, you have to have clearly defined for yourself your abilities or know them so you are an enabling mechanism to help others. It is communal. It is a community development. It's a think tank quite often.

Tyler: I think there's a great history of collaboration and apprenticeship in the paper craft as there has been in all the printing areas. Out of this new interest, I think there is probably going to come a lot of cottage industry that's going to be able to take care of a lot of people, often for a very long time. So, I think there will be apprenticeships opening continually in the paper craft. Whether they will remain in there and become papermakers for a long period of time, I think it's as yet too early to tell.

My suspicion, based on this conference, is that you're going to have now the beginning of a very strong system of apprenticing and collaborating in the field of handmade papermaking. Obviously it's been going on in the machine industry for a very long time.

Noland: I don't think collaboration ever works on an equal basis. When two people work together, one has to help the other, or they can swap. It's like jazz musicians, one can play while the other one backs him up. But it's not equal, somebody has to help someone else.

"Commercialism" and Editioning

Question from Audience: Please discuss the idea of "commercialism" as it relates to editioning versus the unique image.

Tullis: For me, "commercialism" would be making a large edition and then charging large prices for it. There have also been a number of artists who just went and made prints to reproduce paintings, to sell cheap paintings. They didn't look toward the beneficial aspects to society in giving people a quality painting. They just wanted to get paid.

I personally know of major name artists who have sold blank sheets of paper that were signed and numbered and insisted on their check in thousands of dollars before they handed over that gouache to be reproduced in Paris as an original print. Of course, it couldn't be sold in Paris because that's against the law in Paris.

That's commercialism. That's not caring about what you're selling, not having any controls over how your image is made, not drawing the stone yourself, and so on. That's grossly commercial for me. It's unethical.

I'm just saying that if the artist really has care and concern and wants to take a social posture which, if you want to talk about experimentation in art, I think that's where the future of art lies. It is in dealing with many people as opposed to a very elite society. The print has a very powerful political arm in the sense that it can carry unique humanistic ideals in the hands of the right artist to a vast majority of people who are not educated. If you come out of those concerns, you start selling your prints for less, you make them accessible, you facilitate their distribution.

Question from Audience: In assessing these works of art (those from large editions versus uniques), are we changing any of our value system?

Johnson: Yes, one of the things that occurs with the unique object or with the very small edition that is painstakingly printed, or hand painted, or block printed, is a worry that because of the small edition that it will not be judged on its merits but will be misinterpreted and judged as a "precious object." A lithograph or etching, which is published in what I consider reasonable edition of 100 to 150, is something that is going to be judged on its own merits.

But in that very, very small edition of 17 or 22, I have people calling me saying, "I've got one of *those* prints," as if they've got the only one in California. Basically, it's not until about the sixth sentence before they start talking about whether it's a good work of art.

Price and Aesthetics

This causes certain problematical questions as to the aesthetics of the work of art because works of art that are painstakingly printed and worked out of molds that take 82 hours to dry, and take a lot of press time inevitably cost a great deal of money. Works of art that come out of that style have got to be priced accordingly, and possibly that price level is going to cause problems as to whether the work is good.

In the past, when works of art, such as etchings, were $5, $10, or $15, the work of art had to stand or fall on its pure aesthetic merit because it was the type of thing that could be bought or not bought very casually. Today, things are much more difficult. The works of art shown in the museum galleries are usually quite expensive works of art. When I, as a museum curator, consider buying a work of art, I am making a very, very big aesthetic, but also a very big financial commitment.

Tyler: Let me add something else. There is a strong history —both in the dealing of old art and the dealing of new art— that the price level really does exist and that it will be leveled out by the buyers, by the people who want it. I think that in setting the price of unique works or variants of very small editions, one still has to price it with equivalent terms to everything else out there.

In the art profession, as in every profession, there are many people who care a great deal about its survival. I know my conscience comes from talking to people in the museum world, or to scholars, or to dealers for whom I have a great deal of respect, or to the artists who, I feel, have a very keen sense about what price is enough.

I do think, however, that there is a profound danger when something becomes very fashionable, very popular, so that, in a very short period of time, a lot of people get bamboozled. There are a lot of people who could just take advantage of the fact that something is new, current, and exciting to ram the biggest price tag they can down the buyers' throats. I see some of that occasionally on the East Coast since I've been there the last few years. It's more difficult to see it on the West Coast because there is less activity in the art world.

Some of the international movements of the last few years have seen certain artists get unbelievable sums of money for a huge edition of graphics. Now, they cost about one-tenth of what they were asking a few years ago. The market saw to it that the price was dressed down to reality. It gets out there, and somehow it gets corrected. I have a great deal of confidence that the people who buy art and who are engaged in selling it are going to do something to measure it down.

Question from Audience: In making critical appraisals, do you adjust your values in order to accept new experiments in art?

Castleman: I think we'd be hypocritical if we answered that question with any but one answer: If you look at art constantly and constantly make value judgments about it, why should you change your judgments? Why should you if it wants to be considered art?

That's the first thing you have to ask. Do you wish this thing to be considered art? this thing? this object? this piece of paper? whatever? If so, I judge it the same way I judge all art. I don't judge it because it's one of a thousand. I don't judge a book work because anybody can have it for $2. Sure, when it means you have to collect it, you might start thinking in terms of the real world, in dollars and cents and all of that.

But we hope there is a little bit of Parnassus left around, and we all obviously have different judgments. Mine is obviously not always the same as someone else's. It is rarely the same as a lot of other people. But we have to use what we've learned, what we have seen, what we appreciate, where we come from, what we studied, whether we like green or hate blue. All these things come into our judgment.

If something is made out of a material that nobody's been using to its fullest degree for years or for many generations, we should now say, because it's new and different and so forth, we'll just pull out the perimeters of what we accept. I think that art is considered that way. It's a very ethical situation.

Riva Castleman, lecturer and author of numerous books and articles on contemporary printmaking, has juried many major print exhibitions throughout the world. She is currently director of the Department of Prints and Illustrated Books at the Museum of Modern Art, New York.

Robert Flynn Johnson, formerly with the Baltimore Museum of Art, is curator in charge of the Achenbach Foundation for Graphic Arts, Fine Arts Museums of San Francisco.

Kenneth Noland, well-known painter since the late 1950s, has developed his interest in paper as art in the last few years. He has collaborated with Garner and Ann Tullis at the Institute of Experimental Printmaking and with Kenneth Tyler at Tyler Graphics Ltd. to produce some of the most interesting and beautiful paper/art of recent years.

Garner Tullis, artist and teacher, founded the Institute of Experimental Printmaking, in 1973, in Santa Cruz, California. He has encouraged experimentation in his teachings and collaborations with artists who work at the Institute, and has demonstrated this interest in his own artwork. He is currently an assistant professor of art at the University of California at Davis.

Kenneth E. Tyler is the owner of Tyler Graphics Ltd., in Bedford Village, New York.

PAPER AS ART:
A CONTEMPORARY GALLERY

Jules Heller

I would like to begin my presentation with several aphorisms, not attributable to Marshall McLuhan, in one of the languages of the southwestern United States.

No hay arte sin vida;
No hay vida sin desarrollo;
No hay desarrollo sin cambio;
No hay cambio sin controversia;
No hay controversia sin revolucion.

(There is no art without life;
There is no life without development;
There is no development without change;
There is no change without controversy;
There is no controversy without revolution.)

There is little doubt but that handmade papermaking today, especially as used by the artists, is in the midst of a revolution—a technical and aesthetic explosion of artifacts resulting in myriad approaches to visual and plastic expression.

Paper—when employed by artists—may be torn, cut, folded, spindled, or mutilated in many ways. It may be burned, scored, exploded, waxed, dyed, glued, curled, sewn, taped, shaped, stamped, heated, cast, rolled, distressed, crumpled, sawn (as wood), drawn upon, or painted. It may be used as supports for lithographs, intaglios, relief, or screen prints. It may be used for collage, assemblage, frottage, *papier colles,* or three-dimensionally as pulp in molds. It may be recycled again and again, or used as a receptacle, or for embedments. It may be burnished, hammered, or embellished with fibers, feathers, gold and silver leaf, wire, found materials, or mica. It may be made waterproof, fireproof, insectproof, couched (sheet upon sheet), and vacuum-formed. These are but a few of the phenomena you may see in the gallery on the following pages.

There are a number of implied themes that may or may not be evident from the gallery. Here is my listing of them:

1. No matter what anyone says, myself included, there is no one way to get to heaven, or the other place, in papermaking.

2. I will leave the absolute and definitive answer to Rudyard Kipling's question, "It's beautiful, but is it art?"

3. I do not believe in pigeonholing (classifying works according to style, medium, or other categories) the works you see for any purpose. I suggest that pigeonholes are for

pigeons, not for living, breathing workers in the field of papermaking.

4. As a sometime printmaker—one who prints and is familiar with the "black art"—I have often reflected upon the history of papermaking because it is and has always been known as the "white art," regardless of the color of the paper employed. I will return to this notion below.

It is believed that in 1787 Ben Franklin wrote the following words:

"Various the papers, various wants produce,
The wants of fashion, elegance, and use,
Men are as various: and if right I scan,
Each sort of paper represents some man."

With the possible exception of one of Dard Hunter's contributions to papermaking, it is difficult to find, in one volume, the complete story of the so-called "white art" for artist-papermakers from today's vantage point, unless the reader has access to a major library collection of rare and limited edition books on the subject.

Curious. Throughout the centuries, to this very day, people have taken paper for granted. It is regarded as one of the "givens" of society, as ubiquitous as rain, smog, motherhood, or oleomargarine. Being so obvious, it has long been invisible. If requested to think "paper," most individuals will meditate a sheet of white paper. Further, it is widely believed that pure white paper (as with a certain brand of well-advertised soap) is the omega of papermaking.

But, how do you define the color white? What images, what associations come to mind? The albuminous material surrounding the yolk of an egg? the fifth circle of an archery target? the purity and cleanliness of a well-scrubbed white-enamelled kitchen sink? great masses of flour, sugar, and snow?

Snow White and her seven little men? the white part of the eyeball? the silvery-white of the birch? the whiteface of mimes and clowns? whitefish (smoked) for Sunday brunch? the white noise of electronic music?

But enough. Let us leave this intriguing digression to enjoy the following sampling of art works that have used the medium of paper in more than a superficial way.

I would like to thank all the artists who have participated in the selection that follows. I am sorry that, due to the limitations of time and space, this was not able to be a complete survey. There is so much interesting work being done with paper at present that being truly comprehensive is probably unrealizable. My apologies to those who were not included, but we have tried to give an idea of the many creative possibilities that the medium of paper allows.

Finally, I would like to thank Leslie Laird Luebbers, who wrote the captions for all the works presented in the gallery. We would like to thank all of the artists who sent her their comments, which are quoted in the photo captions, as their words help provide valuable insight into how they have worked with the medium.

Jules Heller is dean of the College of Fine Arts and professor of art at Arizona State University at Tempe. Heller is well known as an artist and author of Printmaking Today *and of the recently released* Papermaking, *published by Watson-Guptill. He has traveled throughout the world as visiting professor and has collected extensive knowledge of printmaking and papermaking activities.*

Leslie Laird Luebbers, a printmaker and photographer, is currently research coordinator for the World Print Council.

Neda Al-Hilali
Green Bands 1977
Paper, dye, chalk
36″ × 56″
Courtesy of the Allrich Gallery, San Francisco

Neda Al-Hilali is a fiber artist whose material is paper, and she uses it to the utmost. Beginning with ordinary sheets of paper, she marks on them with traditional pencils, chalk, paints, or dyes. Tradition ends there. She takes the papers apart and re-assembles them by sewing or weaving, and then subjects the work to a new assault of sanding, hammering, compressing, or other unique forms of finishing. The final works are subtle, elegant, and compelling fiber sculptures.

Photo: Gene Krebs

Suzanne Anker
Boundary (Diamond Cutter) 1978
Cast paper, glass, cement, plaster
4′ × 5′
Collection of Mrs. Charles Manassa, St. Louis, Missouri

In 1974, Suzanne Anker turned from the highly technical methods of printmaking to the more direct art-making experience of working with pulp. Anker's method begins with the construction of a plate, which resembles a collograph built up with paper, cardboard, styrofoam, string, and similar materials. This plate is sealed with silicone spray and is then covered with latex to form a negative mold into which the pulp will be cast. If the original plate (and thus the final work) are in high relief, a secondary plaster mold is made to support the latex during the casting and hand-drying process. After casting, the excess water is removed with large sponges, and the piece is allowed to dry. Color may be added to the pulp before casting or applied by hand after the dried work has been removed from the mold. In *Boundary*, as with others in the *Diamond-Cutter* series, Anker has combined the cast paper with cement and glass, thus moving, via papermaking, yet further from traditional printmaking. These works will serve as models for large cast concrete sculptures.

John Babcock
Faulted Arch 1978
Hand-formed paper
32″ × 18″ × 18″
Courtesy of the artist

Making his pulp from cotton linters in a hydropulper of his own invention, Babcock creates paper works that refer in a direct but personal way to the California landscape in which he lives. He explains, "I have always been interested in earth forms and how man responds to this spaceship earth. My most recent work has been layering large slabs of cast papers." *Faulted Arch* is fabricated from 19 sheets. The pulps are colored with various artists' pigments. The surface sheet is mottled, textured, and manipulated while wet to achieve the desired effects.

Laurence Barker
Untitled 1978
Handformed paper with drawing and collage
37″ × 28″
Courtesy of the artist

Laurence Barker studied papermaking with Douglass Howell in the early 1960s, and as head of the printmaking department at Cranbrook in Michigan, he was responsible for providing knowledge and excitement to many artists now engaged in papermaking. Barker currently resides in Spain and continues to combine handmade paper with other media. This work was formed on a screen, and while the pulp was setting, Barker "drew" in it with a stream of water under pressure. He also threw small blobs of pulp into the forming sheet to make depressions and waves. Once dry, the paper was worked with pencil and ink. Finally, pieces of preformed and printed papers were sewn or glued onto the work.

Kati Casida
Whispers III, edition of 2, 1976
Handformed paper, silkscreened with pencil
22″ × 11″
Courtesy of the artist

Kati Casida is a sculptor and printmaker, and her concern with space, rhythm, and surface in both realms of endeavor is clear. She regards printmaking as the intimate, sensual, and particularly tactile one of her crafts, so it seems inevitable that the pulp movement would find a natural and excellent practitioner in Casida. As she says, "Handmade papers offered exactly what I needed for a delicate, but also strong, three-dimensional print or collage. Working with Don Farnsworth, I could regulate the weight of the paper and fold it into shape while the paper was still becoming paper. As with the series *Whispers,* this shaping was rapid, spontaneous and very exciting."

Ronald Davis
Pinwheel, Diamond and Stripe, edition of 42, 1975
Colored intaglio on multicolored paper
20″ × 24″
Courtesy of Tyler Graphics, Bedford, New York

Ron Davis is well known for his paintings and fiberglass works that combine geometric shapes, perspective lines, and richly blended colors, which suggest both the illusion and negation of space. At Tyler Graphics, he collaborated with Ken Tyler to produce paper pieces that deal with the main concerns Davis addresses in other media. They began with a wet, newly made sheet of paper and covered it with a sectioned plastic frame into which they spooned or dabbed pigmented pulp. Finally, the linear element was superimposed by printing over the colored sheet with an intaglio plate.

Photo: David S. Watanabe

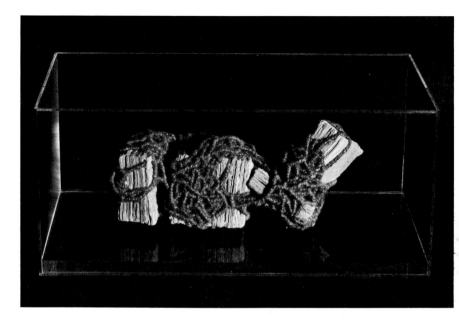

Dominic L. Di Mare
Rune Bundle
Handmade "spun" *gampi* paper, colored pencil, ink, silk, hawthorne wood
8" × 4" × 7"
Courtesy of the artist

The artworks of Dominic L. Di Mare are totally personal in process and product. He makes his papers from natural materials that he finds on his land in northern California. With great delicacy and sensitivity, he combines them with other natural and found objects from his environment. Somehow, working through this process elicits for Di Mare memories of his past childhood spent on his father's fishing boat off California and Mexico. The viewer cannot share that past but can experience personal response to the delicacy and power, as well as the complexity and directness of these artworks created from nature's common gifts.

Tom Fender
Jackson Package 1977
Paper with wool
9¼" × 16¼" × 7¼"
Courtesy of the Allrich Gallery, San Francisco

Tom Fender uses handmade paper in a unique way. He does not use it as a support, which is typical of many of the artists represented in these pages, but, when he uses it as a medium, he presents it in its traditional flat, rectangular, and passive form. The active element in Fender's constructions is the fiber, which writhes, entwines, and finally immobilizes the stacks of paper, which have, however, provided the structure for his work.

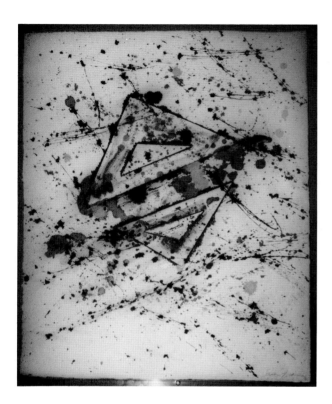

Sam Francis
In, On and Of Paper 1977
Monotype and embossing on handmade paper
30″ × 25″
Courtesy of the Institute of Experimental Printmaking, San Francisco

Sam Francis frequently collaborates on his paperworks with Ann and Garner Tullis, who have managed to develop a unique atmosphere conducive to the creativity of the artist. Sometimes a project will be tackled after a few hours of discussion, sometimes the dialogue continues for months before everyone feels confident to proceed. Francis's paper pieces, like his paintings and lithographs, reveal his uncanny ability to combine spontanteity and structure. For the work illustrated, the metal plate was built up with the triangles, painted and pressed into the damp handmade paper.

Nancy Genn
Marshfield #23 1977
Handmade paper, embossed
38″ × 42″
Courtesy of Susan Caldwell Gallery, New York

Nancy Genn regards paper as both a two-dimensional and as a three-dimensional medium, which draws on her experience as a painter and sculptor. Working in the "Genn Method," which she has perfected over the past three years, Genn lays a wet sheet over previously formed sheets. When each sheet is added and still wet, strips are carefully pulled away to reveal layers beneath. Some of the layers may be pigmented, others may contain natural fibers in addition to the cotton to create color and texture. Finally, the work may be embossed to accentuate or provide contrast to the deckles and torn edges.

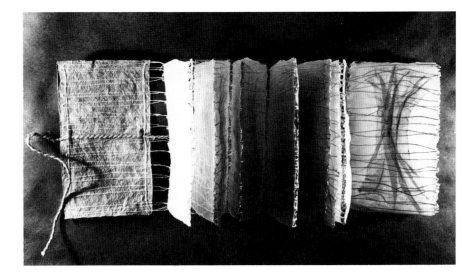

Carolyn Greenwald
Book For Lourdes, one of a series of five books
Handmade Mexican and Oriental papers
9″ × 5″ (closed), 20 pages
Courtesy of the Source Gallery, San Francisco

Carolyn Greenwald's works are concerned with the transparent property of Oriental handmade papers. For the past few years, she has abandoned the notion of paper as support and concentrated on paper as object, while at the same time reducing her works to an intimate scale. Thus, the viewer can manipulate the object and directly participate in the experience of the piece. The book shown here, as with her other works, utilizes only the natural color of the papers. Line is created by imbedding string in the paper, and shadows are made by overlapping folds.

Peter Gentenaar
Transformer 1977-78
Pressed handmade paper and iron strips
5′ × 3′
Courtesy of the artist

In a machine of his own design, Peter Gentenaar vacuum forms flat sheets of colored paper, which he couches on top of each other. The layered wet sheets will adhere to each other in this state, but they are pressed to secure the laminations. Gentenaar is fascinated by the appearance of geologic strata achieved by sawing through the blocks of paper, and he combines these layered forms with sheet metal or iron bars to construct freestanding sculptural plaques. The sense of earthiness is augmented by the naturally oxidized iron. Gentenaar also collaborates with his wife Pat Gentenaar-Torley, a weaver, in making huge molded paperworks. Peter makes a fresh sheet, which Pat then presses into a mold over a textile construction. This will be a unit of the final piece, some being up to three meters square. Peter explains, "Especially in working with textiles, I find that paper becomes very close to textile as a fiber product. But at the same time, it is possible to make it close to a hardboard, which does prove what a nice material it is!"

Clinton Hill
Burst 1977
Handmade paper with watermark and dyed pulp
31″ × 23″
Courtesy of Ellen Sragow, Ltd., New York

Since 1974, Hill has been producing finished works in the process of making paper. The "watermark drawing" is the basis of each piece that Hill makes by attaching bits of plastic or wire to the mold, so that the resulting paper will be thin and transparent in these areas. While the pulp is in the mold, he spoons colored pulp or dye into the sheet. He has frequently worked with papermaker John Koller in Koller's Woodstock, Connecticut, studio.

Shoichi Ida
Paper Between Two Stones and Rock 1978
Lithograph with embossing
37¾″ × 24½″
Courtesy of the Soker-Kaseman Gallery, San Francisco

During the past two years, Ida has produced over 200 works in a series titled *The Surface is the Between.* All the works relate to the paper as surface and substance, as each is worked on both sides of the paper. Often, when fine mulberry paper is used, the inked work on one side interacts with that on the other. Ida will also laminate papers together and peel or scrape the surfaces to cause a subtle texturing. One side may be printed from the floor of the artist's studio, the other with a stone or leaf lithograph. In this way, Ida forces consciousness of the "between," the substance of the paper itself.

Photo: F.L. Avery

Karen Laubhan
New Lullabies Among Ancient Dances, #8 1977
Cotton, linen, sisal, and coconut fibers
36″ × 30″
Courtesy of the artist

Karen Laubhan combined her background in fiberworks and printmaking with experience at the Institute of Experimental Printmaking in Santa Cruz to create her first sculptural paperworks in 1974. She mixes several kinds of fiber into her pulp to achieve texture and strength. The *New Lullabies* series begins with strips of handmade paper, which are dried before being folded, bent, curled, and cemented together with silicone glue.

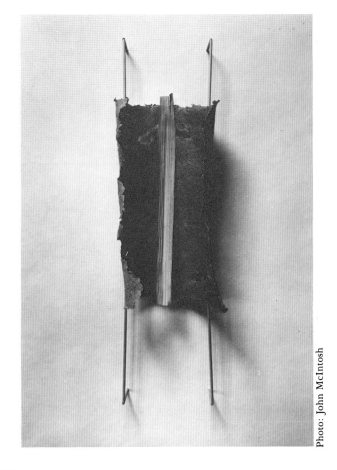

Photo: John McIntosh

Winifred Lutz
Husk 1976
Handmade paper from milkweek pod and seed paper with reed
15¾″ × 23″ × 5″
Collection of Mr. & Mrs. Eugene H. Zagat, Jr.

Like her other works, Winifred Lutz made *Husk* from natural fibers. In her earliest handmade paper, she used sea fibers, such as seaweed. Subsequently, Lutz made paper from many land plants, including pineapple and banana. She is presently interested in making translucent papers from bast fibers such as ramie, abaca, sisal, yucca and flax. According to Lutz, "The paper for *Husk* was formed on a wove mold with an overall watermark of parallel lines. The piece of reed was embossed into it during the first pressing of the post. The curve was established while the paper was still on the blanket, and the paper was left on the blanket to dry to shape."

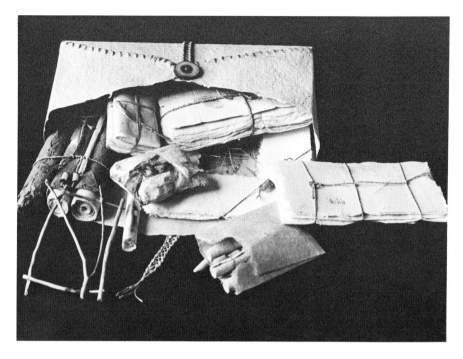

Kenneth Noland
Untitled 1978
Handmade laminated paper
17½ ″ × 23¾ ″
Courtesy of the Institute of Experimental Printmaking, San Francisco

Kenneth Noland first became interested in paper art when he actually got his hands into pulp at the Institute of Experimental Printmaking, now located in San Francisco. Since then, he has collaborated with Ann and Garner Tullis to make his colored pulp pieces. This work required a mold for each color, which represents a layer of pulp built upon the white rectangular base sheet. The colors are bright red, bright yellow, and salmon, and each was achieved by pulping colored rags, not by the addition of pigments. The completed work was pressed to fuse the layers.

Bob Nugent
The Ancient Mariner Series, The Primrose 1978
Handmade rag and mulberry papers, wood, bone, raffia, conte,
19th century English letters, sea urchin spines, hemp
11½ ″ × 15 ″ × 2¾ ″
Courtesy of Grapestake Gallery, San Francisco

Bob Nugent combines contemporary handmade papers with antique papers and natural materials to create an enigmatic package of historical, archaeological, and personal images and discoveries. These "inventions," as Nugent calls them, begin with the actual records of ships sunk off Sable Island in the North Atlantic. Nugent states, "*The Ancient Mariner Series* reflects my fantasies of what one might discover when examining the personal effects, navigational gear, tools, and memorabilia that a ship's captain might take with him on a given voyage. These might include epistles of authority, tokens, love letters, and all of those found and functional objects that contribute to the minutiae of our lives."

Nance O'Banion
Red Flaps and Yellow Triangles 1978
Handmade paper
23″ × 34″
Courtesy of the Allrich Gallery, San Francisco

Nance O'Banion is interested in all the potential of handmade paper. Originally she cast recycled paper, but as she became "seduced" by the abundant possibilities of paper, she began to study traditional papermaking techniques, including that of Eishiro Abe, one of Japan's "National Treasure" papermakers. Her present methods include traditionally pulled and couched rectangular sheets, as well as cast or poured papers. In her own words, "Sometimes both approaches will be used within one piece. Often I will pull a sheet, blow out a grid with pressurized water through a screen, color/draw on it, and then pour it back onto the open grid paper to make a solid piece. After the paper is essentially made and dry, I began to develop the personality with drawing, pastels/crayons, spraying/airbrush through paper stencils with pigments/irridescent acrylics/dyes, sewing lines and even gluing . . . I feel this final metamorphosis gives life/energy/emotion to the papers."

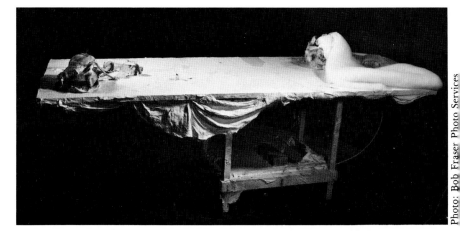

Harold Paris
The Chemite 1978
Paper, cellophane, tape, wood, pins, dacron, glass
40″ × 36″ × 84″
Courtesy of Stephen Wirtz Gallery, San Francisco

Although he is well known for his work in other media, sculptor Harold Paris has been working with handmade paper since 1952, when the medium was introduced to him by Douglass Howell. Since then, he has designed a vacuum pump apparatus, which can be used to form large three-dimensional pieces such as this one. His recent paper works also include smaller, intimate two-dimensional pieces. Always a master of the elusive, Paris offers this comment about his work, derived from the *Annotated Sherlock Holmes,* "When the impossible has been eliminated, the residium, however improbable, must contain the truth."

Michael Ponce de Leon
Succubus, edition of 10, 1967
Collage-intaglio on handmade paper
22½ " × 24 " × 3 "
Collection of Charles Rand Penney Foundation, Alcott, New York

Michael Ponce de Leon has radically expanded the concept of printmaking. In striving to express his personal aesthetics, he has acquired many techniques normally associated with metalsmiths, sculptors, potters, and papermakers. The work that results from this melding of crafts and disciplines could be described as a print only by a miser. They are, in fact, bas-relief intaglio collages on handmade paper. *Succubus* is no exception to Ponce de Leon's complex approach to creation. To produce each two-part print was a five-hour operation, utilizing custom-built molds, handmade inks, and a hydraulic press of his own invention. The spiraling center was cast, in collaboration with Douglass Howell, of linen pulp and linen strips, and printed in 28 colors. The background paper, with the sewer drain, was made with ropes of varying thickness within layers of pulp. Finally, the spiral form was suspended, by means of the linen strips, in front of the background sheet to complete Ponce de Leon's original concept.

Roland Poska
Spring into Fall
Hand-cast paper with pigment and strips of premade paper
6½ ' × 13½ '
Courtesy of Gilman Gallery, Chicago

Roland Poska is a painter and lithographer with many years of experience in making paper by hand. At his Fishy Whale Press in Illinois, Poska casts pulp into huge wooden forms, and while it is still wet, he introduces pigments and slices of paper from the ends of paper slabs, which he has previously made. These slices are called "endforms" and serve to structure his enormous, fresco-like "papestries," which are sometimes exhibited in series to create a gigantic paper environment.

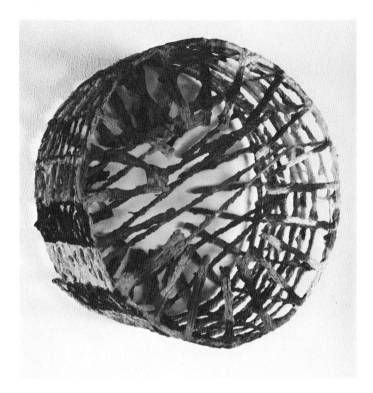

Richard Royce
Frozen Motion Closed 1977
Cast paper print
17″ × 17″ × 16″
Collection of the Portland Museum of Fine Arts

With cast paper, Richard Royce has found the intersection of his concerns with printing and sculpture. According to Royce, "My recent work in cast paper is a logical progression from the work I have done in intaglio from metal plates. I became discouraged when preformed art papers tore and became punctured from the stress of high relief. When I was introduced to the pulp movement in 1975, I realized that casting paper was the solution to my problem because the sheet of paper that is formed at the time of printing did not have stress on it and could take as deep a relief as I wanted. It could conform to curved shapes and go around corners, allowing one to print three-dimensional shapes in the round. I am currently working on a 20-sided spherical structure that is eight feet in diameter."

Alan Shields
Pe Le 1977-78
Wire armature with multiple dips in handmade rag paper with acrylic
9″ × 14″
Courtesy of Paula Cooper Gallery, New York

Alan Shields, one of the most innovative printmakers working today, began several years ago to regard paper as a medium as well as a support. Working in the fertile atmosphere in William Weege's Jones Road Print Shop in Wisconsin, Shields has cut, perforated, folded, interwoven, and constructed lattices with his prints. His handsome paperworks, created in collaboration with Joe Wilfer of the Upper U.S. Paper Mill in Wisconsin, have been combined with prints but also have entered the realm of sculpture with his "dipped" pieces, such as the one illustrated.

Photo: Karen Stahlecker

Karen L. Stahlecker
The Executive Sweets: A Set of Shirts 1978
Handcast paper, pencils and pastels, buttons and sewing thread
Lifesize, installation size about 100″ wide
Courtesy of the artist

"I've been involved with paper since August 1976. For me the medium provided a catalyst for ideas from clay and printmaking." Karen L. Stahlecker also says she has combined the two- and three-dimensional aspects of handmade paper with concepts drawn from a variety of crafts and disciplines. In some cases, as *The Executive Sweets,* she makes sheets of paper upon which she draws or prints in a traditional "art-making" manner and then sews and finishes in a traditional "dress-making" manner. The final forms are exhibited in installations, which suggest group functions of invisible beings in visible clothing. Stahlecker also works in a less representational mode in which she combines seemingly tattered remnants of hand-colored photographs on handmade paper with silk thread and flower petals. She presents the work behind a grid of sticks, which suggests the rectangular format of a photo album.

Photo: Steven Sloman

Frank Stella
Paper Relief—Nowe Miastro, edition of 26, 1975
Dyed and collaged paper with handcoloring
26″ × 21½″
Courtesy of Tyler Graphics, Ltd., Bedford Village, New York

This Frank Stella bas-relief paper piece was made in collaboration with John Koller in Koller's Connecticut studio. Ken Tyler explains the process used to make it, "We made sewn wire moulds shaped in relief from wooden maquettes. These wire moulds were attached to wove moulds and dipped into the vat of pulp and formed like a sheet of paper. This technique presented a problem. Once the pulp was formed over a sculptural mould, the pulp had to be sufficiently dried on the mould before it could be removed. This took roughly 72 hours using a dehumidifier and fans. Stella worked on these papers both in the wet and dry stages with dyes and various other forms of pigmentation. We selected this technique for its sculptural qualities, which relate to the artist's dimensional paintings. The other technique available to us was to make a negative plaster mould and cast pulp into it. However, we liked the idea of actually sewing the wove moulds, which took us four months to do. Also, the paper surface is different in each method, and the artist preferred working on this surface."

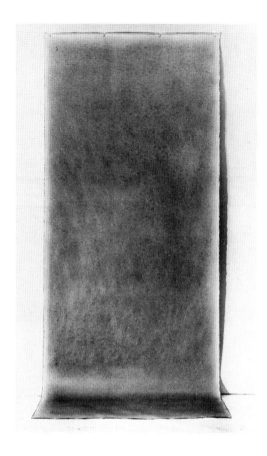

Michelle Stuart
Niagara II #46 1976
Red Queenston shale, natural graphite, all-rag paper mounted on muslin
62′ × 156″
Courtesy of the Museum of Modern Art, Teheran, Iran

Michelle Stuart's works are documents of the time and space of particular natural sites, and the works are done at the site and made from elements found there. Sometimes Stuart creates small books of handmade paper containing the earth or rocks from the site. At other times, working on a monumental scale, she takes a huge roll of paper to the location. Describing the making of *Niagara II*, she writes, "The *Niagara* series was made on the Niagara Gorge. The works are made by violently indenting and impressing the paper repeatedly, all over with rocks and stones from the site, with a larger rock tool, a pestle. The residue from the rocks, the natural earth pigment, is then rubbed and polished in many layers into the surface of the paper."

Garner Tullis
Zabie 1976
Handmade paper and gauze
24″ × 24″ × 10″
Courtesy of the Institute of Experimental Printmaking, San Francisco

Garner Tullis was attracted to pulp as a sculptural medium during the early 1960s. An inveterate experimenter himself, Tullis established the Institute of Experimental Printmaking in Santa Cruz, where many artists and students were helped to explore the vast potential of paper as a medium. Since the relocation of the Institute in San Francisco in 1976, Tullis and his wife Ann concentrate on their personal work and on collaboration with established artists. Garner also teaches at the University of California at Davis. *Zabie* represents one of the ways Tullis works with pulp. He models a clay form, drapes gauze over the form, pats pulp over it, dries it, and removes the gauze and paper sculpture from the form.

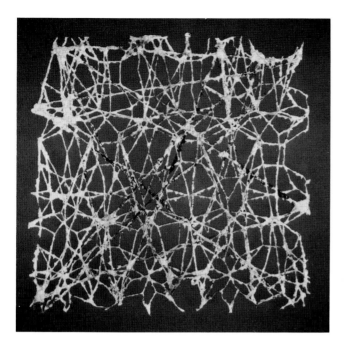

William Weege
San Francisco String Symphony 1977
String, paper pulp, and pigments
4 ′ × 4 ′
Courtesy of the Institute of Experimental Printmaking, San Francisco

William Weege is a printmaker who has always pushed far beyond craft and tradition by combining old and new techniques and materials in revolutionary ways. In his Jones Road Print Shop in Wisconsin, he has not only pursued his own innovations but has facilitated experimentation by other artists. Until recently, handmade paper appeared as just one of many media in his work, but in 1977 at the Institute of Experimental Printmaking in San Francisco, his *String* pieces brought forth pulp as the dominant and dramatic element. These pieces were made by stretching string over a wood frame and dipping it into pulp. After the pulp dried, Weege applied paint, and the work was detached from the frame and pressed.

Joseph Zirker
JZ-SA-37-77 1977
Handformed rag paper with cloth fragments, threads, twine and viscosity printing
15 ″ × 18 ″
Courtesy of the artist

In making his "squashed assemblages," Joseph Zirker begins with a couched sheet of cotton pulp. To this he adds cloth fragments, areas of colored pulp, and linear elements of thread or twine. He then couches a second layer of pulp over the assemblage, which is subsequently squashed in a vacuum table, and then he selectively removes parts of the top layer to reveal the elements below. The work is put back in the vacuum table, resquashed and left to dry. As he did with the illustrated work, Zirker may then print over the finished sheet. He states, "Among my aesthetic concerns has been a focus upon the contrasts that occur when the illusory effects of the viscosity inking are overlaid against the actual dimensional materials of the assemblage."

GLOSSARY

Leslie Laird Luebbers

abaca A Philippine plant, related to the banana, the leaf stalks of which are the source of Manila hemp and a strong, usually buff, paper called Manila paper.

alum A complex salt, most commonly potassium aluminum sulfate, added with rosin to the pulp while it is in the beater as a sizing agent to impart a harder and more water resistant surface to the finished sheet. Alum is also used with dyes as a mordant or color fixitive. Its high acidity causes irreversible damage to works of art.

aquatint A process for producing tonal values on an etching plate. The plate is covered with acid resistant particles (e.g., powdered resin or spray lacquer) and then submerged in acid, which eats away metal between the particles. These bitten areas will hold ink and will produce an even tone when printed. A range of tones can be achieved by "stopping out" or covering some of the aquatinted area with acid resistant varnish and replacing the plate in acid. The aquatint that receives the longer etch will result in a darker tone when printed.

bast fiber The inner bark of such plants as flax, hemp, ramie, *gampi, mitsumata,* and *kōzo,* which serves to distribute food to the plant. When separated from the outer bark, the remaining material provides fibers suitable for making textiles and paper.

beating The process of cutting the fibers to the proper length for sheet formation and bruising or roughing the wall of the fiber, so that more bonding surfaces are available during the papermaking step.

bashu In Japanese papermaking, the stick used for stirring pulp in the vat.

bleaching A continuation of the pulping process by trying to purify the fibers by removing lignins and by whitening. Usually done with chlorine compounds, excessive bleaching may weaken the fibers.

blind print A print made from an uninked plate that has been deeply etched or built up with collage elements. The result is a bas-relief in paper. See embossing.

bonding strength The ability of the fibers at the surface of a paper sheet to adhere to one another and to the fibers below the surface as well as to resist being pulled away, particularly in the process of removing the sheet from a printing plate or stone.

brittleness In paper, the property of cracking or breaking when bent or embossed. This can be the result of overbeating the pulp, which shortens the fibers.

bulk A measurement of the thickness of paper determined by placing a certain number of sheets under a specific pressure.

calendering A process for imparting a smooth, glossy finish to paper by passing it through a series of metal rollers.

caliper An expression of the thickness of one sheet of paper, usually in thousandths of an inch.

cast paper A paper piece, usually three-dimensional, made by pouring pulp into a mold or by patting pulp around a form. The dried paper is then separated from the mold or form and functions on its own as a bas-relief or sculpture.

cellulose A carbohydrate and the chief component of the cell walls of all plants, occurring mainly as long hollow chains called fibers. These fibers have the unique property of adhering together to form a mat from a water suspension (e.g., paper).

chain lines The widely spaced lines in paper made in a laid mold. The lines are created by the impression in the pulp of very thin wires or threads, which are used to sew the numerous, narrowly spaced laid wires to the supporting ribs of the mold.

chemical cotton Commercially purified cotton linter pulp, usually referred to simply as "linters," by papermakers, who may purchase it in various forms as the raw material for paper. See linters.

china paper An extremely smooth and opaque paper achieved by the addition of a fine clay during the beating process. When the sheet is formed, the clay will fill in the interstices of the fiber mat.

chine collé A paper collage process in which sheets of paper are laminated together by the pressure of the etching press and glue. This allows the etcher to achieve colored areas without using a separate plate and also allows the use of papers, which by themselves would be too fragile for the etching process.

close formation An even distribution of fibers throughout a sheet of paper.

collage The use of various materials (e.g., cardboard, metal, plastic, paper) adhered to a surface to create an image.

collograph A relief or intaglio process in which a print is made from a plate built up by collage.

cotton fiber The soft white filaments attached to the seeds of the cotton plant. Cotton fabric is made from long fibers removed from the seeds by "ginning." Short fibers called linters, unsuitable for cloth, but good for papermaking, are left. Paper pulp may be made from cotton rags or linters. Cotton is the purest form of cellulose occurring in nature, so it requires the least processing before it can be used.

couching During the hand-papermaking process, the action of transferring the newly formed sheet from the mold to a dampened felt.

cutting One function of the beating process, which reduces fiber length to the optimum for sheet formation. Long fibers will tend to clump in a "wild formation," while overbeaten fibers will produce a weak, brittle paper.

cylinder paper maker Commonly used today for the production of fine mold-made paper, the cylinder machine was invented at the beginning of the 19th century. A woven metal cylinder rotates, partially suspended in a vat of pulp. A vacuum inside the cylinder hugs the pulp to the screen while sucking out the water, and on the down turn of the cylinder, it transfers the mat of fibers to a continuous felt.

deckle A wooden frame that fits over the mold to prevent the pulp from running off during the dipping and sheet-forming process.

deckle edge The uneven, feathered edge caused by the deckle frame in the production of handmade paper. Originally, this edge was trimmed off, but now it is considered a sign of quality and is desirable.

Edict of Nantes and revocation In 1589, after many years of civil strife in France caused by the political and religious differences between the Catholics and Huguenots (French Calvinists), Henry of Navarre, a Huguenot who converted for political convenience to Catholicism, became King Henry IV of France. In 1598, he issued the Edict of Nantes, which guaranteed civil, religious, and territorial rights to the Huguenots. Nearly a century later, in 1685, Louis XIV tried to display his absolute power over his subjects by revoking the Edict and attempted to force the Huguenots to become Catholics. The most mobile Huguenots, those of the commercial and industrial classes (including many papermakers), fled to Holland, America, and England, taking with them their knowledge, skills, wealth, and ingenuity, which greatly enhanced the economies of those areas at France's expense.

embossing A process that uses a deeply incised or built-up plate to create a raised image. Through the pressure of a press, dampened paper is forced into or around the recessed or raised areas of the plate, resulting in a relief image on the finished print. Embossing may be used without ink (blind print), over ink, or in combination with ink.

engraving An intaglio process in which the image is cut directly into a plate using a tool called a burin or graver.

etching An intaglio process in which an acid-resistant substance is applied to a metal plate. This substance is selectively removed, leaving exposed areas, which will be "etched" or eaten away when the plate is submerged in acid. The etched areas will hold ink to be transferred.

felt 1) A heavy piece of woven cloth, usually wool, upon which the newly formed sheets of paper are couched, or transferred from the mold. 2) The action of removing water from a sheet by hand, with a sponge.

felt finish The surface of a piece of paper resulting from pressing the sheets between pieces of felt. This is the surface of handmade paper, if it is not calendered or sized after the pressing of the "post" (stack of sheets and felts) in order to squeeze out excess water.

fibrillation During the beating process, the fiber walls are bruised and roughened in order to create more surfaces for bonding during sheet formation.

filler In papermaking, materials added to the fiber during the beating stage, which fill the pores of the fiber mat to achieve a harder, whiter, or more opaque surface. Most fillers are minerals such as kaolin clay, calcium carbonate, or titanium dioxide. Filling is also called loading.

flax fiber The bast fiber of the flax plant, which is used in the production of linen and paper. Pulp can be made from the flax fiber or from linen rags.

formation In papermaking, refers to the fiber distribution in a sheet of paper, as it appears when held up to the light. Formation is related to the beating of the pulp and the action of distributing the pulp on the mold.

Fourdriniers The name given to a type of contemporary papermaking machines, which are based on the principle of the continuous web machine invented by Nicholas-Louis Robert in 1798. Several years later, Brian Donkin, an engineer who worked for the London papermakers Henry and Sealy Fourdrinier, improved the machine's design.

fune In Japanese papermaking, the vat which holds the pulp in water suspension for dipping during the sheet forming process. Also called "sukibune."

furnish The combination of ingredients added to the beater in order to make a particular kind of paper. This would include pulped fiber, and any fillers and dyes.

gampi One of the three plants which provide fibers for fine Japanese handmade papers. (Others are *kōzo* and *mitsumata.*) The *gampi* plant grows only in the wild and is decreasing in abundance; therefore, the fiber and the paper are very precious. The paper is very durable, translucent, and nonabsorbent. It is frequently formed on a screen covered with silk cloth, and thus, does not have the chain lines distinguishable in other Japanese papers.

gauffrage print Another term for blind print.

gelatin A colorless protein made from animal tissue and used as an external sizing for paper, so that it will resist staining and bleeding in the printing or coloring process.

glazed paper Sheets of paper, each sandwiched by zinc or slick cardboard plates, are fed through rollers causing a friction of the plates and paper which polishes the surface of the sheets.

gouache A painting medium prepared from opaque water colors and mixed with a gum preparation.

half-stuff Commercially prepared fiber stock, which is purchased in a partially beaten state. Further beating is necessary before the pulp is ready for papermaking.

hemicellulose Cells similar to cellulose, but more subject to destruction by chemicals or atmospheric conditions. However, since they bond more strongly than cellulose, they are desirable for making some papers.

hemp An Asian plant with fibers of high cellulose content, and therefore, suitable for making textiles and paper. Hemp paper is one of the oldest kinds of true paper and was made in China during the first century A.D. Paper made from hemp fiber is strong, white, and lustrous.

Hollander A machine for pulping and refining rags or fibers for papermaking. Invented in Holland at the beginning of the 18th century, this machine replaced the stamping mill. It consists of an oval vat, partially divided in the middle. On one side, a cylinder fitted with blades rotates against a stationary base plate, also fitted with blades. The pulp is propelled around the vat, gradually being refined for the sheet forming process.

hot-pressed paper Paper finished by sandwiching sheets between preheated metal plates and passing them through heavy metal rollers. This results in a smoother surface than regular calendering.

hydration During the beating process, cellulose fibers are bruised and swollen, taking on water. Hydration increases bonding strength but reduces opacity because of the water held in the fiber. In the extreme case, hydration will result in glassine, a dense, translucent paper.

hydropulper A vat with a power-driven agitator/cutter used to reduce and refine fibers to pulp and to "hydrate" the fibers, which increases the bonding potential for papermaking.

intaglio One of the four basic divisions of printmaking (also, lithography, screenprinting, and relief printing). Intaglio includes etching, mezzotint, engraving, drypoint, aquatint, and other processes in which the image is cut below the surface of the plate. Ink is forced into the grooves, the surface wiped clean, and the print is made by the pressure of the press, which forces the dampened paper into the grooves to pick up the ink.

internal sizing Sizing added to the pulp during the beating process.

itaboshi In Japanese papermaking, the process of board drying in sunlight. The side of the paper next to the board becomes the front surface of the paper. This process takes advantage of natural sun bleaching but is vulnerable to weather conditions, which has led to development of drying indoors on heated metal plates.

Japan paper European name for buff-colored, relatively nonabsorbent paper used effectively for printing finely detailed 17th century engravings from Japan.

Jordan refiner A closed, cone-shaped beater or refiner, which allows continuous processing of the pulp, rather than the intermittent beating done in a Hollander.

jute An Indian plant of the linden family. Jute became popular for making paper in Europe in the late 18th century, when the supply of linen and cotton rags failed to meet the demand for paper. The cellulose content of jute is quite high, and it makes a durable paper without extensive refining.

knotter A strainer that removes lumps of fiber from the pulp before it is poured into the vat for sheet forming.

kōzo One of the three principal plants that yield fiber for the production of Japanese paper. (Others are *mitsumata* and *gampi.*) The name is loosely applied to several plants of the mulberry family, which have fibers suitable for papermaking. In general, the long *kōzo* fibers make it the strongest and most dimensionally stable of Japanese papers, and its absorbent surface makes it useful for printing and printmaking.

kurokawa In Japanese papermaking, the black or dark outer bark of the plant after it has been separated from the white inner bark. *Kurokawa* is either discarded or used in the production of "chiri-gami" or "waste paper," until recently used as toilet paper and handkerchiefs but now used for a variety of artistic purposes. It is valued in the West for its natural tan color and fibrous texture.

laid lines The impression or watermark left in paper formed on a laid mold by thin, narrowly spaced wires or bamboo strips spanning the frame in one direction, which support the wet pulp during the papermaking process.

laid mold A wooden frame for the forming of pulp into sheets of paper. The frame is spanned by narrow strips of wire or bamboo, which are held in place by wire or thread worked in the perpendicular direction to the strips.

laminated paper A newly formed sheet of paper is couched on top of another wet sheet instead of felt. The fibers of the sheets will mesh during the drying process, creating layered papers.

layer or **layman** The third person in the hand papermaking trio (others are vat person and coucher), the layer separates the paper from the interleaving felts after the post has been pressed. The sheets are restacked without felts, pressed, parted, restacked, and pressed again, until the desired dryness and smoothness of finish is achieved.

lignin A polymer that is the natural support for the cellulose fibers of woody plants. It rejects water and resists bonding and is, therefore, undesirable in the papermaking process.

linters The short fibers that remain on the cotton seeds after the first "ginning." They are used in the manufacture of paper.

lithography One of the four basic divisions of printmaking. (Others are intaglio, screenprinting, and relief printing.) Planographic, or flat-surface printing, is based on the antipathy of grease and water. An image is made with a greasy substance on a stone or plate, which is then chemically treated so that the image areas take ink and the nonimage areas repell ink. For printing, the stone is linked with a large roller, while the nonimage areas are kept wet, and the image is transferred to paper by the pressure of a press.

lumen The hollow core of the cell wall of a fiber.

marbled paper A decorative paper, frequently used in bookbinding, which simulates the appearance of marble grain. It is made by floating oil based pigments in a vat of water, combing the paint to create a pattern, and then laying paper on the surface of the paint and water to pick up the pigment.

matrix In printmaking, refers to the block, plate, stone, stencil, or other surface on which an image is made for printing. Another term for matrix is "printing element." In commercial printing, it is a mold which is designed to receive positive impressions of type or illustrations from which a negative metal plate can be cast for printing.

mezzotint Also *maniere noire* and *maniere anglaise.* An intaglio process in which the plate surface is uniformly roughened with a tool called a rocker, creating a black background. The artist works from dark to light by scraping and flattening the surface to achieve gray tones and white.

mitsumata One of the three plants (also *gampi* and *kōzo*), which provide materials for Japanese papers. The *mitsumata* fibers are soft and absorbent, and the finished paper has a slight orange color.

mold or **mould** The basic tool of paper-making. In Western papermaking, a wire screen or mesh is attached to a wooden frame. In the Eastern process, a detachable screen is hand held on the wooden frame. In both cases, the mold is dipped into a vat of pulp. The pulp forms a fibrous mat on the screen and the water drains back into the tub. When a sufficient amount of water has drained, the newly formed sheet is couched off the screen, and the mold is ready to make another sheet.

monotype A print that cannot be done in an edition. Frequently, the artists paints an image onto a plate and transfers the image to paper while the paint or ink is still wet, but the term applies to any print process which does not allow duplication.

mordant In printmaking, acid solutions used to bite into an etching plate. In dyeing, a reagent used to fix colors in a material.

naginata beater A Japanese beater similar to the Hollander, in which curved knives mounted on the shaft cut the bast fibers and circulate them around the tub.

neri Vegetable starch, or mucilage, derived from the crushed roots of various plants and added to the vat mixture for Japanese papermaking. This substance, resembling egg white, prevents the fibers from clumping during sheet formation and allows the wet sheets to be separated after they are stacked directly on top of each other and pressed.

parchment A writing or printing surface made from the split hide of sheepskin.

paper A fibrous mat produced by a filtration process in which a dilute slurry of fibers in water is caused to flow across a screen allowing the water to drain out. The sheet is then removed from the screen, pressed, and dried.

papyrus A laminated writing surface made from the sliced stalks of the papyrus plant. The slices are arranged at right angles to each other and pounded together with a mallet. Although not paper, it was the first writing material with properties similar to paper.

post A pile of newly made paper sheets, separated by damp felts, which is ready for pressing to squeeze out excess water.

prepared paper Paper that has been colored by washing pigment over white paper before printing or drawing.

protopaper Any one of several flexible substrates for writing which existed before the invention of paper about A.D. 105, such as palm leaves, papyrus, or silk cloth.

pulp The product of the pulping process, which begins with fibers or rags and liberates or separates them mechanically or chemically. Mechanical pulping simply separates the fibers, while chemical pulping purifies them of lignins and other undesirable agents.

quire A measure of sheets of paper that originally denoted four sheets folded together into eight leaves, but now refers to 24 pieces of paper whether folded or unfolded. A ream is made of 20 quires.

rag content A term describing the amount of cotton fiber relative to the total amount of material used in the making of certain kinds of paper. It is

expressed as a percentage, such as 100% rag content or 80% rag content. The term, though popular, is losing its meaning since more and more high quality paper is made, not from rag, but from linters.

ramie The bast fiber of the ramie plant is used in the production of paper, while the whole stalk is used in making a "wove" mold in parts of China.

ream A quantity of paper sheets, usually 20 quires or 480 sheets, although frequently 500 sheets. The number varies according to the type of paper. A printer's ream is 21½ quires, or 516 sheets.

relief print One of the four basic divisions of printmaking. (Others are lithography, intaglio, and screenprinting.) Relief includes woodcut, linocut, blockprint, etc., in which the image is printed from the surface of the block or plate; areas cut below the surface will not print.

retting Before the use of the Hollander, rags were left to rot on wet stone floors so that the resins in the fibers would weaken sufficiently for the stamper to bruise the fiber.

rice paper An erroneous name for a writing material made from cutting the inner pith of the fatsia plant in a spiral from the outside to the inside. It is not made from rice, and it is not paper, since it is not made from pulped fiber.

rosin One of the most commonly used internal sizing agents for paper. It was first used in the early 1800s, and due to the acidic nature of rosin, papers made since this period usually have a shorter lifespan than papers made before its use.

screenprint A stencil print, and one of the four basic divisions of printmaking. (Others are lithography, intaglio, and relief printing.) A stencil, adhered to a woven screen, blocks out areas of the screen. By pressure of a squeegee, paint or ink is forced through the open spaces onto paper or other surface.

seed-hair fiber Fiber such as cotton, which is attached to the seed of the plant.

seiromushi In Japanese papermaking, the process of steaming plants to soften bark, which is stripped to the white inner bark used for making fine paper. This process is used for *kōzo* and *mitsumata,* but not for *gampi,* which must be stripped at the time of harvesting.

Sekishū-hanshi An unbleached handmade paper made totally from *kōzo* fiber. This paper has been made for over 1,200 years and has been used since then for many purposes, including important documents and account books. It is highly regarded for its beauty, strength, and durability. The production of this paper is protected and subsidized by the Japanese government as an Important Intangible Cultural Property.

sheet formation—Japanese The papermaker dips the mold into the vat and then rocks the mixture back and forth over the surface of the screen. He makes several dips into the vat, gradually adding more layers or laminations, until the desired thickness is achieved. Then the screen or *su* is removed from the frame, and the paper is rolled out onto the stack of previously formed sheets.

sheet formation—Western The papermaker dips the mold and deckle into the slurry and brings it up horizontally in one smooth action. He rocks the mold to even the thickness of the pulp on the screen, and then he gently shakes the mold from side to side and back and forth to "weave" the settling fibers into a mat. The sheet is then transferred to a dampened felt.

shirokawa   In Japanese papermaking, the white, inner bark of plants (usually *kōzo, mitsumata,* or *gampi*) once it has been removed from the outer bark and the intermediate green membrane. It is used to make the most highly regarded handmade papers.

sizing   Additives to the paper fiber, either during the beating stage or after the sheet has been dried, which are intended to control the wetting of the sheet and the ability of the paper to accept ink without feathering.

slurry   The pulp from the beater is added to water in the vat and mixed to achieve the proper suspension for the sheet forming process. The thickness of the sheets will depend on the proportion of pulp to water in the slurry.

stamper   The primitive method of reducing rags and other raw materials to pulp for the papermaking process, the stamper is a mortar and pestle powered by water, wind, or an animal.

stuff   Pulp ready for papermaking. See **half-stuff.**

su   The screen on which the sheet of Japanese paper is formed from the vat. It is made of slender strips of bamboo, held in place by silk thread. It is flexible and can be removed from the mold. The *su* is placed in the hinged mold and deckle made of Japanese cypress.

support   In art, the material (e.g. paper, canvas, plaster, metal) upon which an image is made.

surface sizing   Sizing applied to the dried sheet. Commonly used materials are glue, gelatine, casein, and starch. Also called tub sizing.

trunk fibers   The main wood fiber of trees which provides the chief source of raw material for the commercial paper industry, wood pulp.

vacuum form   In papermaking, pulp is poured over a mold inside an airtight chamber, which is attached by hoses to a compressor and water-collecting tank. When the chamber is closed, the compresor quickly sucks the air and water from the chamber, leaving a partially dried piece of paper in the form of the mold.

vat   A tub that holds the refined pulp mixture into which the mold is dipped for making handmade paper.

vatman or **vat person**   The person who forms the sheets of paper by dipping the mold and shaking the slurry to distribute the fiber.

vellum   A writing surface made from the hide of calf, goat, or lamb, which is treated with lime, scrapped to an even thickness, powdered with chalk, and rubbed with pumice.

viscosity printing   An etching technique that allows multi-color printing from a single plate. The plate is deeply etched and inked with colors, each mixed to a different viscosity (rate of flow) and applied with rollers of varying degrees of hardness. The result is that each color will sit at a different level on the plate and will transfer cleanly to the paper when run through the press.

washi   The Japanese word for the handmade papers traditionally produced in that country.

waterleaf   paper that has no sizing. This paper will absorb fluids readily, and thus is unsuitable for use with thin inks or paints. The term is also applied to a newly made sheet of handmade paper before it is sized.

watermark   A translucent area in a sheet of paper, which is the result of sewing a fine wire design to the surface of the mold screen. In this area, the screen is thicker, and there will be less pulp when the sheet if formed. Papermakers usually watermark their paper with a logo or identifying mark.

wild formation   An uneven distribution of fibers or a clumping of fibers within the pulp of a forming sheet of paper, sometimes caused by excessive fiber length, which is the result of insufficient beating.

wood engraving   A relief print cut from an endgrain block of hardwood (a slice cut at a right angle to the direction of growth). Very fine detail is possible because of the denseness of the fiber and the lack of grain striations.

wove mold   A papermaking form of fine wire mesh, resembling a window screen, which is attached to a wooden frame and is used to make handmade paper.

wove paper   A sheet produced on a wove mold. It has a finer, smoother texture than paper made in a laid mold, and when held to the light, the sheet will reveal a regular, slightly mottled appearance, rather than the horizontal and vertical line pattern of a laid sheet.

xylograph   A wood engraving. In printing, it was an early form of printing text and illustrations from wood blocks, as distinct from the later movable type and metal engraving or etched illustrations.

Leslie Laird Luebbers, a printmaker and photographer, is currently research coordinator for the World Print Council.